Dynamic iteration and model order reduction for magneto-quasistatic systems

Dissertation

zur Erlangung des akademischen Grades

Dr. rer. nat.

eingereicht an der

Mathematisch-Naturwissenschaftlich-Technischen Fakultät

der Universität Augsburg

von

Johanna Kerler-Back

Augsburg, Januar 2019

Universität Augsburg University

Augsburger Schriften zur Mathematik, Physik und Informatik
Band 35

Erstgutachterin: Prof. Dr. Tatjana Stykel
Zweitgutachterin: Prof. Dr. Caren Tischendorf
Tag der mündlichen Prüfung: 14.3.2019

Bibliographic information published by the Deutsche Nationalbibliothek

The Deutsche Nationalbibliothek lists this publication in the Deutsche Nationalbibliografie; detailed bibliographic data are available in the Internet at http://dnb.d-nb.de .

ISBN 978-3-8325-4910-7
ISSN 1611-4256

Logos Verlag Berlin GmbH
Comeniushof, Gubener Str. 47,
10243 Berlin
Tel.: +49 030 42 85 10 90
Fax: +49 030 42 85 10 92
INTERNET: http://www.logos-verlag.de

Danksagung

Zuallererst möchte ich meiner Betreuerin Frau Prof. Dr. Tatjana Stykel herzlich danken. Sie war mir immer eine hilfreiche, geduldige und engagierte Gesprächspartnerin. Ohne ihr stets offenes Ohr wäre diese Arbeit nicht entstanden.

Mein weiterer Dank gilt Frau Prof. Dr. Caren Tischendorf, Herrn Prof. Dr. Timo Reis und Herrn Prof. Dr. Sebastian Schöps für ihre Unterstützung meiner Forschung.

Ich danke meinen Kolleginnen und Kollegen an den Lehrstühlen für Numerische Mathematik und Angewandte Analysis für die angenehme Atmosphäre und die mentale Unterstützung, insbesondere danke ich Frau Dr. Carina Willbold für die anregenden Diskussionen und Herrn Dr. Thomas Fraunholz für die Beantwortung aller meiner Fragen zur Programmierung.

Zuletzt möchte ich noch meiner Familie und Freunden und Freundinnen danken. Sie haben mir in verschiedenster Weise geholfen. Ich danke meinem Mann Matthias Back, der mir immer ein Fels in der Brandung war, meiner Tochter, die mir ein Quell der Freude war, meiner Mutter, die mir stets Mut gemacht hat, meinen Schwiegereltern, meinem Bruder Marian und meiner Schwester Elisabeth, die es mir ermöglicht haben, diese Arbeit zu verfassen ohne meine Tochter zu vernachlässigen, sowie Frau Dr. Isabella Stock, die diese Arbeit Korrektur gelesen hat.

Contents

List of Figures

List of Algorithms

List of Tables

Acronyms and Notation

Acronyms

BT balanced truncation
DAE differential-algebraic equation
DEIM discrete empirical interpolation method
DIRM dynamic iteration using reduced-order models
EVD eigenvalue decomposition
FEM finite element method
FIT finite integration technique
LR-ADI low-rank alternating direction implicit
MDEIM matrix discrete empirical interpolation method
MNA modified nodal analysis
MOR model order reduction
MQS magneto-quasistatic
ODE ordinary differential equation
PDAE partial differential-algebraic equation
PDE partial differential equation
POD proper orthogonal decomposition
SCM successive constraint method
SVD singular value decomposition

Symbols

$\mathbb{C}_+$ open right complex half-plane
$\mathbb{C}_-$ open left complex half-plane
$\lambda_j(A)$ eigenvalues of a matrix A
$\sigma_j(A)$ singular values of a matrix A
$\mathrm{im}(A)$ range of a matrix A
$\ker(A)$ kernel of a matrix A
$\mathrm{trace}(A)$ trace of a matrix A
n state space dimension (also finite element dimension)
m input space dimension
p output space dimension
d space dimension
N number of subsystems in coupled system
η reduced state space dimension (also, POD dimension)
κ DEIM dimension
ρ MDEIM dimension
n_s number of POD snapshots
$\mathcal{C}^k$ space of k times continuously differentiable functions
$\mathcal{H}$ Hilbert space
$\mathcal{L}^p(\Omega)$ Lebesgue space
$\mathcal{W}^{k,p}(\Omega)$ Sobolev space
$\mathcal{H}^k(\Omega) = \mathcal{W}^{k,2}(\Omega)$
$\mathcal{L}^2(0,T;\mathcal{H})$ Bochner space
$\mathcal{H}_\infty$ Hardy space
$\|\cdot\|_2$ spectral matrix norm
$\|\cdot\|_G$ weighted matrix norm
$\|\cdot\|_F$ Frobenius matrix norm
$\|\cdot\|_{\mathcal{H}_\infty}$ Hardy norm
$\|\cdot\|_{\mathcal{L}^2(\Omega)}$ norm in $\mathcal{L}^2(\Omega)$
ξ spatial variable
t time variable
s frequency
u input
x state, often state of the MQS descriptor system
$\tilde{x}$ reduced state, often reduced state of the MQS descriptor system
x_1 state of the MQS standard state space system
$\tilde{x}_1$ POD reduced state of the MQS standard state space system
$\hat{x}_1$ POD-DEIM reduced state of the MQS standard state space system
y output
$\tilde{y}$ POD reduced output
$\hat{y}$ POD-DEIM reduced output
$\hat{f}$ DEIM reduced function
(E, A, B, C, D) DAE system
$(\tilde{E}, \tilde{A}, \tilde{B}, \tilde{C}, \tilde{D})$ reduced DAE system
$(\mathcal{E}, \mathcal{A}, \mathcal{B}, \mathcal{C})$ MQS descriptor system
$(\mathcal{E}_r, \mathcal{A}_r, \mathcal{B}_r, \mathcal{C}_r)$ regular DAE system for 3D MQS model
$(\mathcal{E}_1, \mathcal{A}_1, \mathcal{B}_1, \mathcal{C}_1)$ MQS standard state space system
$(\tilde{\mathcal{E}}_1, \tilde{\mathcal{A}}_1, \tilde{\mathcal{B}}_1, \tilde{\mathcal{C}}_1)$ reduced MQS standard state space system
$\mathbf{G}$ transfer function
$\tilde{\mathbf{G}}$ approximated transfer function
G_c controllability Gramian
G_o observability Gramian
ς_j Hankel singular values

Electromagnetics

Ω bounded domain
Ω_1 conducting bounded domain
Ω_2 non-conducting bounded domain
σ electric conductivity
ϵ electric permitivity
μ magnetic permeability
ν magnetic reluctivity
$\mathbf{E}$ electric field density
$\mathbf{D}$ electric flux density
$\mathbf{H}$ magnetic field density
$\mathbf{B}$ magnetic flux density
φ electric scalar potential
$\mathbf{A}$ magnetic vector potential
$\mathbf{A}_0$ initial data for magnetic vector potential
χ_{str} winding function for stranded conductors
Υ factor of coupling function for stranded conductors
$\mathbf{J}_s$ external source current density
$\mathbf{J}_c$ conduction current density
$\boldsymbol{a}$ discrete potential
$\mathcal{K}$ stiffness matrix in the MQS system
$\mathcal{M}$ mass matrix in the MQS system
$\mathcal{X}$ coupling matrix in the MQS system
m_{ν} monotonicity constant of magnetic reluctivity
$\mathcal{R}$ restistance in stranded coil
v voltage in coil
ι current in coil

1 Introduction

Our world today is becoming more and more complex. Technical applications such as mobile phones, computers, industrial machines, cars and even coffee machines and watches are becoming more compact and more powerful. An essential aspect in the development of new semiconductor devices including processors and memory chips is the ongoing miniaturization of structures on the one hand and the continuous increase of working frequencies on the other side. The high density of electronic components together with high clock frequencies leads to unwanted side-effects like crosstalk, delayed signals and substrate noise, which are no longer negligible in chip design and can only insufficiently be represented by simple lumped circuits models. As a result, different physical phenomena such as electromagnetic induction and heat transfer have to be taken into consideration since they have an increasing influence on the signal propagation in the circuits. Due to the extremely high costs of prototype fabrication and testing newly developed electronic devices, methods of computer-based simulation play a key role in the design and production process.

The modelling and analysis of complex multi-physics problems typically leads to coupled systems of partial differential equations (PDEs) and differential-algebraic equations (DAEs). DAE systems are used to model dynamic processes subject to certain constraints resulting from physical laws like conservation laws and Kirchhoff's network laws. PDEs are used to describe spatial or spatio-temporal phenomena in different subcomponents. Application examples can be found in various fields, e.g., in the analysis of electronic and electromagnetic circuits, micro-electro-mechanical systems and fluid-structure interactions. The high dimension of the spatially discretized PDEs and the heterogeneity of system components require the development of new powerful algorithms which, on the one hand, allow a fast simulation under consideration of the modular system structure and, on the other hand, enable higher accuracy and robustness of the numerical results.

Co-simulation or the coupling of different simulation tools for the analysis of coupled systems has already been used for simulation of flow-structure interactions in the 1980s, see [LBH84], and has become a standard technique for the analysis of complex multidisciplinary models in nonlinear system dynamics over the past thirty years [FPF01, GTB$^+$18, SLL15, WP99]. In co-simulation, the individual subsystems are iteratively solved by adjusting the coupling conditions by means of dynamic iteration. Thereby the combination of different integration methods (multi-method approach) and the use of different time steps (multi-rate methods) for the individual model components is possible. Stability and convergence of the dynamic iteration depend on many parameters: type of used iteration method, number of iterations, macro step size and formulation of the coupling conditions. Though for special system structures, stability and convergence can often be ensured, this does not apply to more general structures [AG01, Ebe08, PT18, SA12].

With the increasing complexity of multiphysics problems to be modelled, the interest in model order reduction has grown rapidly in recent years [BHtM11, BMS05, BOA$^+$17, SvdVR08]. The aim of model reduction is to approximate high-dimensional systems with models of lower dimension which capture the input-output behaviour of the original system as accurately as possible, preserve important physical properties such as stability and passivity and require much less simulation time. For linear (DAE) systems, various model reduction techniques have been developed and successfully applied in numerous application areas such as modal approximation in structural mechanics [Dav66, Mar66], Krylov subspace-based methods in electromagnetic and microsystem technology [Bai02, Fre04, OCP98] and balanced truncation in circuit simulation and fluid dynamics [RS10, RS11, HSS08]. Several model reduction techniques have also been developed for nonlinear systems. These include the trajectory piecewise linear method [RW06], methods based on bilinear and quadratic-linear approximations [BB12] and the proper orthogonal decomposition (POD) approach [Vol99] combined with the discrete empirical interpolation method (DEIM) [CS10]. Model reduction of linear coupled systems has been considered in [RS07], where error bounds for reduced coupled system were derived which depend on the error bounds for the subsystems.

In this thesis, we first investigate the dynamic iteration using reduced-order models (DIRM) which was proposed in [RP02]. In numerical experiments, we observed that this method depends on many parameters that strongly influence its convergence. Our goal is to derive an error estimator which provides a reliable information on accuracy of the computed solution at low cost. Such an error estimator could further be used in the convergence analysis of the DIRM method which still remains an open problem.

In the second part of this thesis, we study coupled field-circuit systems. Assuming that the contribution of the displacement currents is negligible compared to the conductive currents and that the conductivity vanishes on a non-conducting subdomain, the magnetic field can be described by magneto-quasistatic (MQS) systems which can be considered as an approximation to Maxwell's equations. Such systems are used for modeling of low-frequency electromagnetic devices. A spatial discretization of coupled MQS-circuit systems leads to large-scale DAEs whose numerical solution requires an enormous amount of storage and large computational time. To reduce numerical effort, model order reduction can be used. Since model reduction of circuit equations has already been well studied, see, e.g., [OCP98, Fre04, RS10, RS11], we will focus here on MQS systems. Model reduction of electromagnetic problems is currently a very active research area [HC13, HC14, KS15, MHCG16, NST14, NT13] because faster simulations are essential in parameter study and computational optimization of electromagnetic structures. For model reduction of linear MQS systems, we employ the balanced truncation approach, whereas nonlinear MQS systems are reduced using the POD-DEIM method. We will exploit the special block structure of the underlying problem to improve the performance of the model reduction algorithms.

Outline of the thesis

This thesis is organized as follows. In Chapter 2, we present a basic concepts for DAEs and linear control systems. We also introduce common model reduction methods for linear and nonlinear systems and discuss the numerical solution of PDEs using the finite element method (FEM). In Chapter 3, we consider the numerical solution of coupled systems. In Section 3.1, we first describe dynamic iteration scheme which is based on splitting the time interval into macro time windows and solving the subsystems iteratively on each time window. In Section 3.2, dynamic iteration is combined with the POD model reduction approach providing the DIRM method. The performance of this method for different choice of method parameters will then be investigated by using a simple coupled system in Section 3.3. Finally, in Section 3.4, we derive an a posteriori error estimator for the DIRM method and discuss its computation. Numerical experiments demonstrate the accuracy properties of the DIRM error estimator. The preliminary results of this section have been presented in [KS14].

In Chapter 4, we consider model reduction of linear and nonlinear MQS systems. In Section 4.1, we introduce general Maxwell's equations describing the electric and magnetic fields and derive a MQS approximation to these equations in Section 4.2. In Section 4.3, we investigate first the 2D MQS system. Based on a weak formulation, a spatial discretization of such a system using FEM is discussed. We analyze the structural and physical properties of the resulting DAE. In particular, we show that the semidiscretized MQS system is of tractability index one and passive. A transformation of this DAE system into the ordinary differential equation (ODE) form is also presented. Furthermore, in Section 4.4, we develop a balanced truncation model reduction method for 2D linear MQS systems. This method is based on solving one Lyapunov matrix equation and preserves passivity. Section 4.5 deals with model reduction of 2D nonlinear MQS systems using the POD method combined with DEIM for efficient evaluation of the nonlinearity. We prove that the POD reduced model is passive and present a passivity enforcing method for the POD-DEIM reduced model. Some parts of Sections 4.4 and 4.5 have been published in [KBS17]. In Section 4.6, we report some results of numerical experiments for a single-phase 2D transformer model. Additionally, we compare the numerical solutions of a coupled MQS-circuit system obtained by four different numerical methods: monolithic approach, dynamic iteration, monolithic approach combined with model reduction of the MQS subsystem, and dynamic iteration combined with model reduction. Finally, in Section 4.7, we extend several results for 2D MQS systems to 3D models. To overcome the difficulty caused by singularity of the 3D semidiscretized MQS system, we present a regularization approach which is based on projecting out the singular variables and ensures the unique solvability of the regularized system.

2 Preliminaries

In this chapter, we briefly review differential-algebraic equations and linear control systems and discuss their properties. We also introduce model reduction methods for linear and nonlinear systems and outline the finite element method for solving elliptic partial differential equations.

2.1 Differential-algebraic equations

First, we collect some facts on differential-algebraic equations (DAEs). More details can be found in [HW96, KM06]. DAEs arise in different applications such as electrical circuits, multi-body systems, computational fluid dynamics and electromagnetic problems. Some of these applications will also be considered in Chapter 4.

Consider an implicit differential equation

$$0 = F(\dot{x}, x, t), \tag{2.1.1}$$

where $F : \mathbb{R}^n \times \mathbb{R}^n \times \mathbb{I} \to \mathbb{R}^n$, $\mathbb{I} = (t_0, T) \subset \mathbb{R}$, $x : \mathbb{I} \to \mathbb{R}^n$ is a continuously differentiable function, and $\dot{x}$ denotes the derivative of x with respect to $t \in \mathbb{I}$. Equation (2.1.1) together with an initial condition

$$x(t_0) = x_0 \tag{2.1.2}$$

forms the initial value problem. A function $x : \mathbb{I} \to \mathbb{R}^n$ is called a *solution* of the initial value problem (2.1.1), (2.1.2) if x is continuously differentiable and satisfies (2.1.1) pointwise and also (2.1.2). If the initial value problem (2.1.1), (2.1.2) is solvable, then the initial condition (2.1.2) is called *consistent*.

If $\frac{\partial F}{\partial \dot{x}}$ exists and is invertible, then system (2.1.1) can be solved with respect to $\dot{x}$. In this case, the implicit differential equation (2.1.1) is equivalent to a system of ordinary differential equations (ODEs). Here, we consider (2.1.1), where $\frac{\partial F}{\partial \dot{x}}$ is singular. Such a system contains the differential and (possibly hidden) algebraic equations and is called a system of DAEs. The most simple form of DAEs is the linear time-invariant DAE system given by

$$E\dot{x} = Ax + f(t), \quad x(t_0) = x_0, \tag{2.1.3}$$

where $E, A \in \mathbb{R}^{n \times n}$, $f : \mathbb{R} \to \mathbb{R}^n$ and $x_0 \in \mathbb{R}^n$. For this system, we have

$$F(\dot{x}, x, t) = E\dot{x} - Ax - f(t)$$

and, therefore,

$$\frac{\partial F}{\partial \dot{x}}(\dot{x}, x, t) = E.$$

Thus, for E invertible, (2.1.3) is an ODE, and for E singular, we have a DAE. The existence and uniqueness of solution of such a system can be studied by considering the eigenvalue structure of the matrix pencil $\lambda E - A$.

Definition 2.1. For a pencil $\lambda E - A$, $\lambda \in \mathbb{C}$ is called a *finite eigenvalue* and $v \in \mathbb{R}^n$ is called an *eigenvector* if $Av = \lambda Ev$. The pencil $\lambda E - A$ has an *eigenvalue at infinity* if $\mu = 0$ is the eigenvalue of the pencil $E - \mu A$.

Assume that the pencil $\lambda E - A$ is *regular*, i.e., $\det(\lambda E - A) \neq 0$ for some $\lambda \in \mathbb{C}$. Then $\lambda E - A$ can be transformed into the *Weierstraß canonical form*

$$E = T_l \begin{bmatrix} I_{n_f} & 0 \\ 0 & E_\infty \end{bmatrix} T_r, \qquad A = T_l \begin{bmatrix} A_f & 0 \\ 0 & I_{n_\infty} \end{bmatrix} T_r, \tag{2.1.4}$$

where T_r and T_l are nonsingular transformation matrices, $n_f + n_\infty = n$, E_∞ is a nilpotent matrix with *index of nilpotency* k, and A_f contains the finite eigenvalues of the pencil. Let

$$T_r x = \begin{bmatrix} x_1 \\ x_2 \end{bmatrix}, \qquad T_l^{-1} f = \begin{bmatrix} f_1 \\ f_2 \end{bmatrix}, \qquad T_r x_0 = \begin{bmatrix} x_{01} \\ x_{02} \end{bmatrix} \tag{2.1.5}$$

be partitioned accordingly to E and A in (2.1.4). Then the DAE system (2.1.3) can be written as

$$\dot{x}_1 = A_f x_1 + f_1, \quad x_1(t_0) = x_{01}, \tag{2.1.6a}$$

$$E_\infty \dot{x}_2 = x_2 + f_2, \quad x_2(t_0) = x_{02}. \tag{2.1.6b}$$

Equation (2.1.6a) is a linear ODE. It is uniquely solvable for all initial vectors $x_{01} \in \mathbb{R}^{n_f}$. Equation (2.1.6b) has also a unique solution given by

$$x_2 = -\sum_{i=0}^{k-1} E_\infty^i \frac{d^i f_2}{dt^i}$$

provided f_2 is sufficiently smooth and the initial vector $x_{02} \in \mathbb{R}^{n_\infty}$ satisfies the consistency condition

$$x_{02} = -\sum_{i=0}^{k-1} E_\infty^i \frac{d^i f_2}{dt^i}(t_0).$$

This shows that unlike ODEs, for the existence of the continuously differentiable solution x of the DAE (2.1.3), it is necessary that the inhomogeneity f is k times continuously differentiable and x_0 belongs to the set of consistent initial conditions

$$\mathcal{X}_0 = \left\{ T_r^{-1} \begin{bmatrix} x_{01} \\ x_{02} \end{bmatrix} \;:\; x_{01} \in \mathbb{R}^{n_f},\ x_{02} = -\sum_{i=0}^{k-1} E_\infty^i \frac{d^i f_2}{dt^i}(t_0) \right\}.$$

To classify how close a DAE (2.1.1) is being to an ODE, different index concepts were introduced in the literature [HW96, KM06, LMT13]. An overview can be found in [Meh15, Sch03b]. First, we define the differentiation index following [Cam87]. The main idea thereby is to find out how often one has to differentiate the DAE system (2.1.1) in order to reformulate it as an ODE system.

Definition 2.2. Equation (2.1.1) with a sufficiently smooth F has a *differentiation index* δ if δ is the minimal number of differentiations

$$\begin{aligned} 0 &= F(\dot{x}, x, t), \\ 0 &= \frac{dF(\dot{x}, x, t)}{dt}, \\ &\vdots \\ 0 &= \frac{d^{\delta} F(\dot{x}, x, t)}{dt^{\delta}} \end{aligned} \tag{2.1.7}$$

such that (2.1.7) allows to extract by algebraic manipulations an ODE system

$$\dot{x} = \varphi(x, t), \tag{2.1.8}$$

which is called *underlying ODE*.

Note that the underlying ODE (2.1.8) has the same dimension as the DAE (2.1.1). We now apply this definition to the linear DAE (2.1.3). This system has the differentiation index $\delta = 1$ if

$$\begin{aligned} E\dot{x} &= Ax + f(t), \\ E\ddot{x} &= A\dot{x} + \dot{f}(t) \end{aligned} \tag{2.1.9}$$

is solvable for $\dot{x}$ such that $\dot{x}$ depends only on x and t but not on $\ddot{x}$. In order to determine the underlying ODE, we consider two matrices Y_l and Y_r whose columns form the bases of $\ker(E)$ and $\ker(E^T)$, respectively. Extend Y_l and Y_r with Z_l and Z_r, respectively, such that the matrices $\left[Z_l, Y_l\right]$ and $\left[Z_r, Y_r\right]$ are invertible. Multiplying both equations in (2.1.9) from the left with $\left[Z_l, Y_l\right]^T$ and introducing the new variables

$$\begin{bmatrix} x_1 \\ x_2 \end{bmatrix} = \begin{bmatrix} Z_r & Y_r \end{bmatrix}^{-1} x,$$

we obtain

$$\begin{aligned} Z_l^T E Z_r \dot{x}_1 &= Z_l^T A Z_r x_1 + Z_l^T A Y_r x_2 + Z_l^T f(t), \\ 0 &= Y_l^T A Z_r x_1 + Y_l^T A Y_r x_2 + Y_l^T f(t), \\ Z_l^T E Z_r \ddot{x}_1 &= Z_l^T A Z_r \dot{x}_1 + Z_l^T A Y_r \dot{x}_2 + Z_l^T \dot{f}(t), \\ 0 &= Y_l^T A Z_r \dot{x}_1 + Y_l^T A Y_r \dot{x}_2 + Y_l^T \dot{f}(t). \end{aligned} \tag{2.1.10}$$

We know that

$$\operatorname{rank}(Z_l^T E Z_r) = \operatorname{rank}\left(\begin{bmatrix} Z_l^T E Z_r & 0 \\ 0 & 0 \end{bmatrix}\right) = \operatorname{rank}\left(\begin{bmatrix} Z_l^T \\ Y_l^T \end{bmatrix} E \begin{bmatrix} Z_r & Y_r \end{bmatrix}\right) = \operatorname{rank}(E).$$

Therefore, the matrix $Z_l^T E Z_r \in \mathbb{R}^{\operatorname{rank}(E) \times \operatorname{rank}(E)}$ is invertible and we can solve the first equation in (2.1.10) for $\dot{x}_1$. If $Y_l^T A Y_r$ is invertible, we can solve the last equation in (2.1.10) for $\dot{x}_2$. Then the underlying ODE has the form

$$\begin{aligned} \dot{x}_1 &= (Z_l^T E Z_r)^{-1} \left(Z_l^T A Z_r x_1 + Z_l^T A Y_r x_2 + Z_l^T f(t)\right), \\ \dot{x}_2 &= -\left(Y_l^T A Y_r\right)^{-1} Y_l^T A Z_r (Z_l^T E Z_r)^{-1} \left(Z_l^T A Z_r x_1 + Z_l^T A Y_r x_2 + Z_l^T f(t)\right) \\ &\quad - \left(Y_l^T A Y_r\right)^{-1} Y_l^T \dot{f}(t). \end{aligned}$$

In this case, the DAE (2.1.3) has the differentiation index $\delta = 1$. The invertibility of $Y_l^T A Y_r$ is equivalent to the condition $\operatorname{rank}[E, AY_r] = n$. If $Y_l^T A Y_r$ is not invertible, then system (2.1.3) has at least the differentiation index $\delta = 2$.

Next, we introduce a tractability index from [LMT13]. The idea is now to avoid the differentiation of the equations and work instead with projectors. For simplicity, we restrict ourselves to a quasilinear DAE

$$F\frac{d}{dt}d(x,t) + b(x,t) = 0 \tag{2.1.11}$$

where $F \in \mathbb{R}^{n\times l}$ and $d : \mathbb{R}^n \times \mathbb{I} \to \mathbb{R}^l$ and $b : \mathbb{R}^n \times \mathbb{I} \to \mathbb{R}^n$ are sufficiently smooth functions. We only consider DAEs, where $D(x,t) := \frac{d}{dx}d(x,t)$ has a constant rank. In addition, we demand a properly stated leading term property

$$\ker(F) \oplus \operatorname{im}(D(x,t)) = \mathbb{R}^n, \quad x \in \mathbb{R}^n, t \in \mathbb{I}.$$

Definition 2.3 ([Tis03, Definition A.14]). A quasilinear DAE (2.1.11) with properly stated leading term is said to be *regular with tractability index* τ if there exist two sequences of continuous matrix-valued functions

$$\begin{aligned}\mathcal{G}_0(x,t) &= FD(x,t),\\ \mathcal{G}_{i+1}(x^i,\dots,x^1,x,t) &= \mathcal{G}_i(x^{i-1},\dots,x^1,x,t) + \mathcal{B}_i(x^i,\dots,x^1,x,t)\mathcal{Q}_i(x^{i-1},\dots,x^1,x,t),\end{aligned}$$

and

$$\begin{aligned}\mathcal{B}_0(x,t) &= \frac{\partial b}{\partial x}(x,t),\\ \mathcal{B}_{i+1}(x^{i+1},\dots,x^1,x,t) &= \mathcal{B}_i(x^i,\dots,x^1,x,t)\mathcal{P}_i(x^{i-1},\dots,x^1,x,t)\\ &\quad -(\mathcal{G}_{i+1}D^-\mathrm{Diff}_{i+1})(x^{i+1},\dots,x^1,x,t)(D\mathcal{P}_0\cdots\mathcal{P}_i)(x^{i-1},\dots,x^1,x,t),\end{aligned}$$

with

$$\begin{aligned}\mathrm{Diff}_1(x^1,x,t) &= \frac{\partial(D\mathcal{P}_0\mathcal{P}_1D^-)}{\partial x}(x,t)x^1 + \frac{\partial(D\mathcal{P}_0\mathcal{P}_1D^-)}{\partial t}(x,t),\\ \mathrm{Diff}_{i+1}(x^{i+1},\dots,x^1,x,t) &= \sum_{j=1}^{i}\frac{\partial(D\mathcal{P}_0\cdots\mathcal{P}_{i+1}D^-)(x^i,\dots,x^1,x,t)}{\partial x^j}x^{i+1}\\ &\quad + \frac{\partial(D\mathcal{P}_0\cdots\mathcal{P}_{i+1}D^-)(x^i,\dots,x^1,x,t)}{\partial x}x^1\\ &\quad + \frac{\partial(D\mathcal{P}_0\cdots\mathcal{P}_{i+1}D^-)(x^i,\dots,x^1,x,t)}{\partial t},\end{aligned}$$

where for all $x^i,\dots,x^1,x \in \mathbb{R}^n$ and $t \in \mathbb{I}$,

1. $\mathcal{Q}_i(x^{i-1},\dots,x^1,x,t)$ is a continuous projector function onto the subspace $\ker(\mathcal{G}_i(x^{i-1},\dots,x^1,x,t))$,
2. $\mathcal{P}_i(x^{i-1},\dots,x^1,x,t) = I - \mathcal{Q}_i(x^{i-1},\dots,x^1,x,t)$,
3. $\mathcal{Q}_i(x^{i-1},\dots,x^1,x,t)\mathcal{Q}_j(x^{j-1},\dots,x^1,x,t) = 0$ for $j = 0,\dots,i-1$, $i > 0$,
4. D^- is a reflexive generalized inverse of D satisfying $DD^- = R$, $D^-D = \mathcal{Q}_0$ and DD^- is a projector onto $\operatorname{im}(D(x,t))$,
5. $D\mathcal{P}_0\cdots,\mathcal{P}_iD^-(x^{i-1},\dots,x^1,x,t)$ is continuously differentiable,
6. $\mathcal{G}_i(x^{i-1},\dots,x^1,x,t)$ has constant rank $r_i > 0$,
7. $r_{\tau-1} < r_\tau = n$.

In order to demonstrate the application of this definition to the linear DAE system (2.1.3), we transform first this system into a system with properly stated leading term. Let the columns of Z form a basis of $\operatorname{im}(E)$. Then system (2.1.3) can be written in the form of (2.1.11) with $F = EZ(Z^TZ)^{-1}$, $d(x,t) = Z^Tx$ and $b(x,t) = -Ax - f(t)$. We have

$$\begin{aligned}\mathcal{G}_0 &= EZ(Z^TZ)^{-1}Z^T = E,\\ \mathcal{B}_0 &= A,\end{aligned}$$

as $Z(Z^TZ)^{-1}Z^T$ is a projector onto $\operatorname{im}(E)$. If $\mathcal{G}_0$ is invertible, system (2.1.3) is equivalent to the ODE

$$\dot{x} = E^{-1}Ax + E^{-1}f(t).$$

If $\mathcal{G}_0$ is singular, we find $\mathcal{Q}_0$ as a projector onto $\ker(\mathcal{G}_0)$ and $\mathcal{P}_0 = I - \mathcal{Q}_0$. Then we have

$$\mathcal{G}_1 = \mathcal{G}_0 + \mathcal{B}_0\mathcal{Q}_0 = E + A\mathcal{Q}_0.$$

If $\mathcal{G}_1$ is invertible, then the DAE (2.1.3) has the tractability index $\tau = 1$.

Note that for the linear DAE (2.1.3) with a regular pencil $\lambda E - A$, the differentiation and tractability indices coincide and are equal to the nilpotency index of $\lambda E - A$. However, for nonlinear DAEs different indices characterize various properties of the system. For other index concepts such as strangeness index, pertubation index, geometric index, structural index and relation between them, we refer to [HW96, KM06, Pan88, Rhe84] and [Meh15], respectively.

2.2 Linear control systems

In this section, we present some control-theoretic concepts for a linear time-invariant control system

$$\begin{aligned}E\dot{x} &= Ax + Bu,\\ y &= Cx + Du,\end{aligned} \tag{2.2.1}$$

where $x \in \mathbb{R}^n$ is the state, $u \in \mathbb{R}^m$ is the input, $y \in \mathbb{R}^p$ is the output, $E, A \in \mathbb{R}^{n\times n}$, $B \in \mathbb{R}^{n\times m}$, $C \in \mathbb{R}^{p\times n}$ and $D \in \mathbb{R}^{p\times m}$. If $E = I$, system (2.2.1) is called *standard state space system.* If E is nonsingular, system (2.2.1) can easily be transformed into the standard state space form by multiplication from the left with E^{-1}. In the case when E is singular, system (2.2.1) is called *descriptor system* or *generalized state space system.*

2.2.1 Standard state space systems

First, we consider a standard state space system

$$\begin{aligned}\dot{x} &= Ax + Bu,\\ y &= Cx + Du.\end{aligned} \tag{2.2.2}$$

As we are often interested only in the input-output behavior, we introduce the transfer function and list some of its properties.

Definition 2.4. The matrix-valued function

$$\mathbf{G}(s) = C(sI - A)^{-1}B + D$$

is called the *transfer function* of system (2.2.2).

The transfer function $\mathbf{G}$ describes the input-output behavior of systen (2.2.2) in the frequency domain. This follows from the relation $\mathbf{y}(s) = \mathbf{G}(s)\mathbf{u}(s)$ between the Laplace transformed input $\mathbf{u}$ and the Laplace-transformed output $\mathbf{y}$ which is obtained by applying the Laplace transformation to system (2.2.2).

Definition 2.5. A quadruple (A, B, C, D) is called a *state space realization* of a transfer function $\mathbf{G}$ if

$$\mathbf{G}(s) = C(sI - A)^{-1}B + D.$$

For every invertible matrix T, the quadruple $(TAT^{-1}, TB, CT^{-1}, D)$ gives another realization of $\mathbf{G}(s) = C(sI - A)^{-1}B + D$.

Definition 2.6. The Hardy norm $\|F\|_{\mathcal{H}_\infty}$ of a function $F : \mathbb{C} \to \mathbb{C}^{q\times r}$, which is analytic in the right complex half-plane $\mathbb{C}_+ = \{z \in \mathbb{C} : \operatorname{Re}(z) > 0\}$, is defined by

$$\|F\|_{\mathcal{H}_\infty} = \sup_{z\in\mathbb{C}_+} \sigma_{\max}(F(z)),$$

where $\sigma_{\max}(F(z))$ denotes the largest singular value of the matrix $F(z)$.

One can show that if the transfer function $\mathbf{G}$ of system (2.2.2) is analytic in $\mathbb{C}_+$, then

$$\|\mathbf{G}\|_{\mathcal{H}_\infty} = \sup_{u\neq 0} \frac{\|y\|_{\mathcal{L}^2(0,\infty;\mathbb{R}^p)}}{\|u\|_{\mathcal{L}^2(0,\infty;\mathbb{R}^m)}},$$

where $\|f\|_{\mathcal{L}^2(0,\infty;\mathbb{R}^m)}$ denotes the $\mathcal{L}^2$-norm of a function $f : (0,\infty) \to \mathbb{R}^m$. Thus, the $\mathcal{H}_\infty$-norm is the induced norm to the $\mathcal{L}^2$-norm. This gives directly

$$\|y\|_{\mathcal{L}^2(0,\infty;\mathbb{R}^p)} \leqslant \|\mathbf{G}\|_{\mathcal{H}_\infty} \|u\|_{\mathcal{L}^2(0,\infty;\mathbb{R}^m)}.$$

Note that $\mathbf{G}$ is analytic in $\mathbb{C}_+$ if all eigenvalues of A have negative real part. In this case, any solution of the homogeneous equation $\dot{x} = Ax$ satisfies $\lim_{t\to\infty} x(t) = 0$. This property is known as *asymptotic stability* [HNW93, Chapter I.13].

Next, we introduce the concepts of controllability and observability for system (2.2.2).

Definition 2.7. A control system (2.2.2) is called *controllable*, if for any initial condition $x(t_0) = x_0$, $t_1 > t_0$ and a final state $x_1 \in \mathbb{R}^n$, there exists a control $u : [t_0, t_1] \to \mathbb{R}^m$ such that the solution of the state equation in (2.2.2) satisfies $x(t_1) = x_1$.

Definition 2.8. A control system (2.2.2) is called *observable* if for any $t_1 > t_0$, the initial state $x(t_0) = x_0$ can be determined from the time history of the input $u(t)$ and the output $y(t)$ on the interval $[t_0, t_1]$.

The controllability and observability properties of (2.2.2) can be characterized by the observability and controllability Gramians defined as follows.

Definition 2.9. Let system (2.2.2) be asymptotically stable, i.e., all eigenvalues of A have negative real part. Then the *controllability* and *observability* Gramians of (2.2.2) are defined by

$$\begin{aligned} G_c &= \int_0^\infty e^{A\tau} B B^T e^{A^T\tau} d\tau, \\ G_o &= \int_0^\infty e^{A^T\tau} C^T C e^{A\tau} d\tau. \end{aligned} \tag{2.2.3}$$

The following theorem shows that the Gramians can be determined by solving the Lyapunov matrix equations.

Theorem 2.10. *Let system* (2.2.2) *be asymptotically stable. Then the controllability and observability Gramians satisfy the Lyapunov equations*

$$\begin{aligned} AG_c + G_c A^T &= -\, BB^T, \\ A^T G_o + G_o A &= -\, C^T C. \end{aligned}$$

Proof. See [Ant05, Proposition 4.27]. □

Note that for system (2.2.1) with nonsingular E, the controllability and observability Gramians solve the *generalized Lyapunov equations*

$$AG_cE^T + EG_cA^T = -\, BB^T, \tag{2.2.4}$$
$$A^TG_oE + E^TG_oA = -\, C^TC. \tag{2.2.5}$$

They exist if all eigenvalues of the pencil $\lambda E - A$ have negative real part.

2.2.2 Descriptor systems

We now consider a descriptor system given by

$$\begin{aligned} E\dot{x} &= Ax + Bu, \\ y &= Cx + Du, \end{aligned} \tag{2.2.6}$$

where $E \in \mathbb{R}^{n\times n}$ is singular. Similar to the standard state space case, we define the transfer function for (2.2.6) as follows.

Definition 2.11. The matrix-valued function

$$\mathbf{G}(s) = C(sE - A)^{-1}B + D$$

is called the *transfer function* of the descriptor system (2.2.6).

Definition 2.12. A transfer function $\mathbf{G}$ of (2.2.6) is called *proper* if $\lim_{s\to\infty} \mathbf{G}(s)$ exists and *improper* otherwise. If $\lim_{s\to\infty} \mathbf{G}(s) = 0$, then $\mathbf{G}$ is called *strictly proper*.

Consider the Weierstraß canonical form (2.1.4) for the pencil $\lambda E - A$ with the transformation matrices T_l and T_r and introduce

$$T_r x = \begin{bmatrix} x_1 \\ x_2 \end{bmatrix}, \quad T_l^{-1} B = \begin{bmatrix} B_1 \\ B_2 \end{bmatrix}, \quad C T_r^{-1} = \begin{bmatrix} C_1 & C_2 \end{bmatrix}.$$

Then the descriptor system (2.2.6) can be decoupled into a slow subsystem

$$\begin{aligned} \dot{x}_1 &= A_f x_1 + B_1 u, \\ y_1 &= C_1 x_1, \end{aligned} \tag{2.2.7}$$

and a fast subsystem

$$\begin{aligned} E_\infty \dot{x}_2 &= x_2 + B_2 u, \\ y_2 &= C_2 x_2 + D u \end{aligned} \tag{2.2.8}$$

with $y = y_1 + y_2$. Analogously, we can split the transfer function

$$\mathbf{G}(s) = \mathbf{G}_{sp}(s) + \mathbf{G}_p(s) \tag{2.2.9}$$

into a strictly proper part

$$\mathbf{G}_{sp}(s) = C_1(sI - A_f^{-1})B_1$$

and a polynomial part

$$\mathbf{G}_p(s) = C_2(sE_\infty - I)^{-1} B_2 + D = \sum_{j=0}^{k-1} M_j s^j$$

with

$$\begin{aligned} M_0 &= -C_2 B_2 + D, \\ M_j &= -C_2 E_\infty^j B_2, \quad j = 1, \dots, k-1. \end{aligned}$$

In order to extend the controllability and observability Gramians to descriptor systems, we need the spectral projectors

$$P_l = T_l \begin{bmatrix} I_{n_f} & 0 \\ 0 & 0 \end{bmatrix} T_l^{-1}, \qquad P_r = T_r^{-1} \begin{bmatrix} I_{n_f} & 0 \\ 0 & 0 \end{bmatrix} T_r$$

onto the left and right deflating subspaces of the pencil $\lambda E - A$ corresponding to the finite eigenvalues and the complementary projectors

$$Q_l = I - P_l, \qquad Q_r = I - P_r,$$

which are the spectral projectors onto the left and right deflating subspaces corresponding to the eigenvalue at infinity. These projectors allow us to define the controllability and observability Gramians for the slow and fast subsystems separately.

We first define the *proper controllability* and *observability Gramians* G_{pc} and G_{po} of system (2.2.6) as unique symmetric positive semidefinite solutions of the *projected continuous-time Lyapunov equations*

$$\begin{aligned} EG_{pc}A^T + AG_{pc}E^T &= -P_lBB^TP_l^T, & G_{pc} &= P_rG_{pc}P_r^T \\ E^TG_{po}A + A^TG_{po}E &= -P_r^TC^TCP_r, & G_{po} &= P_l^TG_{po}P_l. \end{aligned}$$

Such Gramians exist if system (2.2.6) is asymptotically stable meaning that all finite eigenvalues of the pencil $\lambda E - A$ have negative real part. Furthermore, the *improper controllability* and *observability Gramians* G_{ic} and G_{io} are defined as unique symmetric positive semidefinite solutions of the *projected discrete-time Lyapunov equations*

$$\begin{aligned} AG_{ic}A^T - EG_{ic}E^T &= Q_lBB^TQ_l^T, & G_{ic} &= Q_rG_{ic}Q_r^T, \\ A^TG_{io}A - E^TG_{io}E &= Q_r^TC^TCQ_r, & G_{io} &= Q_l^TG_{io}Q_l. \end{aligned}$$

Finally, the controllability and observability Gramians of the descriptor system (2.2.6) are defined as $G_c = G_{pc} + G_{ic}$ and $G_o = G_{po} + G_{io}$, respectively. Unlike standard state space systems, there exist different types of controllability and observability for descriptor systems [BRT17]. These system properties can be characterized by special rank conditions for the proper and improper Gramians, see [Sty04, Corollary 2.5] for details.

2.3 Passivity

In this section, we introduce passivity. Passive systems form a special class of dissipative dynamical systems which have extensively been studied in [HM80, vdS00, Wil72]. They are of particular interest in circuit simulation [AV73] and controller design [BLME07, CS05]. Roughly speaking, a system is passive if it does not generate energy or, equivalently, the energy dissipates. Mathematically, passivity can be defined in terms of a storage function or an available storage characterizing the maximum energy that can be extracted from the system. An important property of passive systems is that an interconnection of passive subsystems often provides a new passive system [Wil72].

For our later purposes, we define passivity for a semilinear autonomous descriptor system

$$\begin{aligned} E\dot{x} &= f(x,u), \quad x(0) = x_0, \\ y &= g(x,u) \end{aligned} \tag{2.3.1}$$

where $E \in \mathbb{R}^{n\times n}$, $f : \mathbb{R}^n \times \mathbb{R}^m \to \mathbb{R}^n$, $h : \mathbb{R}^n \times \mathbb{R}^m \to \mathbb{R}^m$ and $x_0 \in \mathbb{R}^n$. We know that if E is singular, the initial vector x_0 should satisfy the consistency condition depending on u. On the other side, for given initial condition $x(0) = x_0$, we can consider only those inputs which are *admissible* with this condition meaning that the initial value problem (2.3.1) is solvable.

Definition 2.13. A descriptor system (2.3.1) is called *input-output passive* (io-passive) if for all $T > 0$ and all squared integrable inputs $u : [0,T] \to \mathbb{R}^m$ admissible with the initial condition $x(0) = 0$, the corresponding output $y : [0,T] \to \mathbb{R}^m$

satisfies

$$\int_0^T y^T(\tau)u(\tau)d\tau \geqslant 0.$$

For a linear descriptor system (2.2.6), io-passive can be characterized in terms of positive realness of the transfer function $\mathbf{G}(s) = C(sE - A)^{-1}B + D$. The transfer function $\mathbf{G}$ is called *positive real* if $\mathbf{G}$ is analytic in the right complex half-plane $\mathbb{C}_+$ and $\mathbf{G}(s) + \mathbf{G}^*(s) \geqslant 0$ for all $s \in \mathbb{C}_+$. The following theorem shows an equivalence between io-passivity and positive realness of $\mathbf{G}$.

Theorem 2.14. *A linear descriptor system* (2.2.6) *is io-passive if and only if its transfer function* $\mathbf{G}(s) = C(sE - A)^{-1}B + D$ *is positive real.*

Proof. See [AV73]. □

The next theorem provides sufficient conditions for the linear descriptor system (2.2.6) to be io-passive.

Theorem 2.15. *A descriptor system* (2.2.6) *with* $\ker(E) \cap \ker(A) = \emptyset$ *and*

$$E = E^T \geqslant 0, \quad A + A^T \leqslant 0, \quad B = C^T, \quad D = 0 \tag{2.3.2}$$

is io-passive.

Proof. First, we show that all finite eigenvalues of the pencil $\lambda E - A$ lie in the closed left half-plane. This implies that $\mathbf{G}$ is analytic in $\mathbb{C}_+$, since the poles of $\mathbf{G}$ are the eigenvalues of $\lambda E - A$. Let λ_0 be a finite eigenvalue of $\lambda E - A$ and let v be the corresponding eigenvector. Then $Ev \neq 0$, $\lambda_0 Ev = Av$ and $\overline{\lambda}_0 v^* E^T = v^* A^T$. Moreover, we have

$$0 \geqslant v^*(A + A^T)v = \lambda_0 v^* Ev + \overline{\lambda}_0 v^* E^T v = 2\mathrm{Re}(\lambda_0) v^* Ev.$$

Since $v^* Ev > 0$, we obtain $\mathrm{Re}(\lambda_0) \leqslant 0$. Using (2.3.2), we obtain that

$$\begin{aligned}\mathbf{G}(s) + \mathbf{G}^*(s) &= C(sE - A)^{-1}B + B^T(\overline{s}E^T - A^T)^{-1}C^T \\ &= C(sE - A)^{-1}(2\mathrm{Re}(s)E - (A + A^T))^{-1}(\overline{s}E^T - A^T)^{-1}C^T \geqslant 0\end{aligned}$$

for all $s \in \mathbb{C}_+$. Hence, by Theorem 2.14 system (2.2.6) is io-passive. □

There is another definition of passivity which we also introduce here.

Definition 2.16. A descriptor system (2.3.1) is said to be *passive* if there exists a nonnegative function $S : \mathbb{R}^n \to \mathbb{R}_0^+$, such that $S(0) = 0$ and for all $T \geqslant 0$ and all quadratically integrable inputs u admissible with an initial condition $x(0) = x_0 \in \mathbb{R}^n$, the passivation inequality

$$S(x(T)) - S(x_0) \leqslant \int_0^T y^T(\tau)u(\tau)\, d\tau \tag{2.3.3}$$

holds, where x solves the initial value problem (2.3.1) on the time interval $[0, T]$. The function S is called the *storage function.*

Remark 2.17. Note that passivity of (2.3.1) immediately implies io-passivity. Indeed, assume that system (2.3.1) is passive. Then for all $T > 0$ and all inputs u admissible with the initial condition $x(0) = 0$, it follows from Definition 2.13 that

$$\int_0^T y^T(t)u(t)\,dt \geqslant S(x(T)) - S(x(0)) = S(x(T)) \geqslant 0.$$

Thus, (2.3.1) is io-passive. For standard state space nonlinear systems, it has been shown in [HM80, Pol98] that under an additional assumption of reachability, Definitions 2.13 and 2.16 are equivalent. This result was extended to linear descriptor systems in [Brü10]. Passivity of infinite-dimensional descriptor systems was studied in [JR08].

2.4 Model order reduction

In this section, we outline model order reduction (MOR) of linear and nonlinear systems using balanced truncation and proper orthogonal decomposition, respectively. In case of nonlinear systems, we also present a discrete empirical interpolation method for fast evaluation of nonlinear function. Consider a nonlinear control system

$$\begin{aligned} E\dot{x} &= f(x,u), \\ y &= g(x,u). \end{aligned} \tag{2.4.1}$$

where $E \in \mathbb{R}^{n\times n}$, $f : \mathbb{R}^n \times \mathbb{R}^m \to \mathbb{R}^n$ and $g : \mathbb{R}^n \times \mathbb{R}^m \to \mathbb{R}^p$. We assume that the dimension n of the state x is very large, while the dimensions m and p of the input u and the output y, respectively, are small compared to n. The goal of MOR is to compute a reduced model

$$\begin{aligned} \tilde{E}\dot{\tilde{x}} &= \tilde{f}(\tilde{x},u), \\ \tilde{y} &= \tilde{g}(\tilde{x},u), \end{aligned}$$

which has a reduced state space dimension $\eta \ll n$ and the approximation error $\|y - \tilde{y}\|$ is small in some appropriate function norm. A common model reduction approach is reduction by projection. Here, we are searching for two projection matrices $V \in \mathbb{R}^{n\times\eta}$ and $W \in \mathbb{R}^{n\times\eta}$. Substituting an approximation $x \approx V\tilde{x}$ into (2.4.1), we get the residual $r = EV\dot{\tilde{x}} - f(V\tilde{x},u)$. If W is chosen such that $W^T r = 0$, we obtain the reduced-oder model

$$\begin{aligned} W^T EV\dot{\tilde{x}} &= W^T f(V\tilde{x},u), \\ \tilde{y} &= g(V\tilde{x},u). \end{aligned}$$

To gain a speed up in the construction of the reduced-order models and their simulation, we split the computations in a computationally expensive offline stage and a cheap online stage.

2.4.1 Balanced truncation for linear systems

For linear systems, there are different methods to generate the projection matrices W and V, see [ASG01, BBF14] for an overview. In this section, we present a balanced truncation (BT) model reduction approach introduced first in [Moo81] and further studied in [Enn84, LHPW87, TP87]. This model reduction technique strongly relies on the control-theoretic concepts such as controllability and observability Gramians introduced in Section 2.2. A main idea of BT is to transform the system into a balanced form and then truncate the states that are difficult to control and to observe.

Balanced truncation for standard state space systems

First, we present the BT method for the linear standard state space system (2.2.2). For such a system, the reduced-order model has the form

$$\begin{aligned} \dot{\tilde{x}} &= \tilde{A}\tilde{x} + \tilde{B}u, \\ \tilde{y} &= \tilde{C}\tilde{x} + \tilde{D}u, \end{aligned} \tag{2.4.2}$$

where

$$\tilde{A} = W^T AV, \qquad \tilde{B} = W^T B, \qquad \tilde{C} = CV, \qquad \tilde{D} = D \tag{2.4.3}$$

with appropriately chosen projection matrices W and V. For linear systems, the reduced-order matrices (2.4.3) can be precomputed in the offline stage and only a linear system of dimension η has to be solved in the online stage.

In oder to be able to measure the importance of states, we introduce two energy functions

$$\begin{aligned} \mathcal{E}_u &= \|u\|^2_{\mathcal{L}^2(-\infty,0;\,\mathbb{R}^m)} = \int_{-\infty}^0 u^T(t)u(t)dt, \\ \mathcal{E}_y &= \|y\|^2_{\mathcal{L}^2(0,\infty;\,\mathbb{R}^p)} = \int_0^{\infty} y^T(t)y(t)dt. \end{aligned}$$

The input energy $\mathcal{E}_u$ is the square of $\mathcal{L}^2$-norm of the input signal on the time interval $(-\infty, 0)$, whereas the output energy $\mathcal{E}_y$ is the square of the $\mathcal{L}^2$-norm of the output signal on the time interval $(0, \infty)$.

The following theorem shows a connection between the energy functions and the controllability and observability Gramians.

Theorem 2.18. *Assume that system* (2.2.2) *is asymptotically stable. Let* G_c *and* G_o *be the controllability and observability Gramians as defined in* (2.2.3).

1. *If system* (2.2.2) *is controllable, then* G_c *is nonsingular and*

$$\min_{\substack{u\in\mathcal{L}^2(-\infty,0;\,\mathbb{R}^m),\\ x(-\infty)=0,\\ x(0)=x_0}} \mathcal{E}_u = x_0^T G_c^{-1} x_0$$

is the minimal energy required to steer the state $x(-\infty) = 0$ *to* $x(0) = x_0$.

2. *The output energy produced by system (2.2.2) with an initial condition $x(0) = x_0$ and $u(t) = 0$ for $t \geqslant 0$ is given by*

$$\mathcal{E}_y = x_0^T G_o x_0.$$

Proof. See [Ant05, Lemma 4.29]. □

The importance of the state variables is measured using the *Hankel singular values* defined as

$$\sigma_j = \sqrt{\lambda_j(G_c G_o)},$$

where $\lambda_j(G_c G_o)$ denote the eigenvalues of $G_c G_o$. They are assumed to be ordered decreasingly. One can show, see [Ant05, Lemma 5.8], that the Hankel singular values σ_j are the singular values of the *Hankel operator*

$$\begin{aligned} \mathcal{H} : \mathcal{L}^2(-\infty, 0; \mathbb{R}^m) &\to \mathcal{L}^2(0, \infty; \mathbb{R}^p), \\ u(t) &\mapsto y(t) \end{aligned}$$

defined by

$$y(t) = (\mathcal{H}u)(t) = \int_{-\infty}^{0} Ce^{A(t-\tau)} Bu(\tau) d\tau, \quad t \geqslant 0.$$

Let $G_c = Z_c Z_c^T$ and $G_o = Z_o Z_o^T$ be the Cholesky factorizations of the Gramians. Then the Hankel singular values can be determined as the singular values of the matrix $Z_o^T Z_c$. This follows from the relation

$$\sigma_j^2 = \lambda_j(G_c G_o) = \lambda_j(Z_c Z_c^T Z_o Z_o^T) = \lambda_j(Z_c^T Z_o Z_o^T Z_c) = \sigma_j^2(Z_o^T Z_c),$$

where $\lambda_j(\cdot)$ and $\sigma_j(\cdot)$ denote the j-th eigenvalue and singular value, respectively.

System (2.2.2) is called *balanced* if the controllability and observability Gramians are equal and diagonal with the Hankel singular values on the diagonal, i.e., $G_c = G_o = \operatorname{diag}(\sigma_1, \ldots, \sigma_n)$. Consider the singular value decomposition (SVD) $Z_o^T Z_c = U \Sigma Q^T$, where U und Q are orthogonal and $\Sigma = \operatorname{diag}(\sigma_1, \ldots, \sigma_n)$. If system (2.2.2) is controllable and observable, then (2.2.2) can be transformed into the balanced form $(T_b A T_b^{-1}, T_b B, C T_b^{-1}, D)$ with the balancing transformation $T_b = \Sigma^{-\frac{1}{2}} U^T Z_o^T$ and $T_b^{-1} = Z_c Q \Sigma^{-\frac{1}{2}}$.

We now take a look at the energy functions for the balanced system. Since $G_c = G_o = \Sigma$, the energy functions read as

$$\min_{\substack{u \in \mathcal{L}^2(-\infty, 0; \mathbb{R}^m), \\ x(-\infty) = 0, \\ x(0) = e_j}} \mathcal{E}_u = e_j^T G_c^{-1} e_j = \frac{1}{\sigma_j},$$

and

$$\mathcal{E}_y = e_j^T G_o e_j = \sigma_j,$$

where e_j denotes the j-th column of the identity matrix. This means that for large σ_j, the states are easy to control and observe. The truncation is performed by splitting

Algorithm 2.1: Balanced truncation for standard state space systems

Input : an asymptotically stable system (A, B, C, D)
Output: a reduced-order asymptotically stable system $(\tilde{A}, \tilde{B}, \tilde{C}, \tilde{D})$

1 Solve the Lyapunov equations

$$AG_c + G_c A^T = -BB^T,$$
$$A^T G_o + G_o A = -C^T C,$$

for the Cholesky factorizations $G_c = Z_c Z_c^T$ and $G_o = Z_o Z_o^T$.

2 Compute the SVD

$$Z_o^T Z_c = \begin{bmatrix} U_1 & U_0 \end{bmatrix} \begin{bmatrix} \Sigma_1 & \\ & \Sigma_0 \end{bmatrix} \begin{bmatrix} Q_1 & Q_0 \end{bmatrix}^T,$$

where $[U_1, U_0]$ and $[Q_1, Q_0]$ are orthogonal, $\Sigma_1 = \operatorname{diag}(\sigma_1, \ldots, \sigma_\eta)$ and $\Sigma_0 = \operatorname{diag}(\sigma_{\eta+1}, \ldots, \sigma_n)$.

3 Compute the projection matrices $W = Z_o U_1 \Sigma_1^{-\frac{1}{2}}$ and $V = Z_c Q_1 \Sigma_1^{-\frac{1}{2}}$.

4 Compute the reduced matrices $\tilde{A} = W^T AV$, $\tilde{B} = W^T B$, $\tilde{C} = CV$ and $\tilde{D} = D$.

$U = [U_1, U_0]$, $Q = [Q_1, Q_0]$ and $\Sigma = \operatorname{diag}(\Sigma_1, \Sigma_0)$ and taking the projection matrices $W = Z_o U_1 \Sigma_1^{-\frac{1}{2}}$ and $V = Z_c Q_1 \Sigma_1^{-\frac{1}{2}}$. We summarize the BT method in Algorithm 2.1.

In the following theorem, we collect the properties of the reduced-order model.

Theorem 2.19. *Let system* (2.2.2) *be asymptotically stable and let*

$$\mathbf{G}(s) = C(sI - A)^{-1}B + D$$

be its transfer function. Furthermore, let (2.4.2) *be a reduced-order model computed by Algorithm 2.1. Then*

1. *the reduced system* (2.4.2) *is asymptotically stable and balanced,*

2. *for the transfer function* $\tilde{\mathbf{G}}(s) = \tilde{C}(sI - \tilde{A})^{-1}\tilde{B} + \tilde{D}$ *of* (2.4.2), *we have the error bound*

$$\|\tilde{\mathbf{G}} - \mathbf{G}\|_{\mathcal{H}_\infty} \leqslant 2(\sigma_{\eta+1} + \ldots + \sigma_n).$$

Proof. The asymptotic stability of the reduced-order system has been proven in [PS82]. The error bound has been derived in [Enn84, Glo84]. □

Remark 2.20. The BT method can also be rewritten for system (2.2.1) with a nonsingular matrix E. To this end, in Algorithm 2.1, the Lyapunov equations in Step 1 should be replaced by the generalized Lyapunov equations (2.2.4) and (2.2.5) and the SVD of $Z_o^T Z_c$ in Step 2 by the SVD of $Z_o^T E Z_c$. The computation of the reduced matrices $\tilde{A}$, $\tilde{B}$, $\tilde{C}$ and $\tilde{D}$ remain the same and for $\tilde{E}$, it applies that

$$\begin{aligned}\tilde{E} &= W^T EV = \Sigma_1^{-\frac{1}{2}} U_1^T Z_o^T E Z_c Q_1 \Sigma_1^{-\frac{1}{2}} \\ &= \Sigma_1^{-\frac{1}{2}} U_1^T \begin{bmatrix} U_1 & U_0 \end{bmatrix} \begin{bmatrix} \Sigma_1 & \\ & \Sigma_0 \end{bmatrix} \begin{bmatrix} Q_1 & Q_0 \end{bmatrix}^T Q_1 \Sigma_1^{-\frac{1}{2}} = I.\end{aligned}$$

Low-rank alternating direction implicit method for Lyapunov equations

We now briefly discuss the numerical solution of the generalized Lyapunov equations (2.2.4) and (2.2.5). For this purpose, we use the low-rank alternating direction implicit (LR-ADI) method. It was first introduced in [LW02, Pen00] and enhanced in [BKS13a, BKS13b]. In many applications, it was observed that the solution of the Lyapunov equations (2.2.4) and (2.2.5) with a low-rank right-hand side has a low numerical rank and, hence, it can be well approximated by a low-rank matrix $G_c \approx \tilde{Z}\tilde{Z}^T$, where $\tilde{Z} \in \mathbb{R}^{n\times r}$, $r \ll n$, is called low-rank Cholesky factor of G_c. Such a factor can be computed by the LR-ADI iteration given by

$$\begin{aligned} F_k &= (\tau_k E + A)^{-1} Y_{k-1}, \\ Y_k &= Y_{k-1} - 2\mathrm{Re}(\tau_k) E F_k, \\ Z_k &= \begin{bmatrix} Z_{k-1} & \sqrt{-2\mathrm{Re}(\tau_k)} F_k \end{bmatrix} \end{aligned} \tag{2.4.4}$$

with initial matrices

$$\begin{aligned} F_1 &= (\tau_1 E + A)^{-1} B, \\ Y_1 &= B - 2\mathrm{Re}(\tau_1) E F_1, \\ Z_1 &= \sqrt{-2\mathrm{Re}(\tau_1)} F_1 \end{aligned}$$

and shift parameters $\tau_k \in \mathbb{C}_-$. As a stopping criterion, we use

$$\|R\|_F \leqslant tol \|B^T B\|_F$$

with the residual

$$R = A Z_k Z_k^T E^T + E Z_k Z_k^T A^T + B B^T = Y_k Y_k^T$$

and a given tolerance $tol > 0$. The shift parameters τ_k are chosen in conjugated pairs to keep Y_k and Z_k real, see [BKS14] for more details. They strongly influence the convergence of the LR-ADI iteration. In [Pen00], it was proposed to use the smallest and largest Ritz values of the pencil $\lambda E - A$ which can be computed by an Arnoldi algorithm [Arn51]. Other strategies for computing the LR-ADI shifts have been considered in [BKS14, GSA03, Wac13].

Balanced truncation for descriptor systems

We now present an extension of the BT model reduction approach to the descriptor system (2.2.6) proposed first in [LS01, Sty04]. Our goal is to find a reduced-order system

$$\begin{aligned} \tilde{E}\dot{\tilde{x}} &= \tilde{A}\tilde{x} + \tilde{B}u, \\ \tilde{y} &= \tilde{C}\tilde{x} + \tilde{D}u, \end{aligned} \tag{2.4.5}$$

where $\tilde{E} = W^T E V$, $\tilde{A} = W^T A V$, $\tilde{B} = W^T B$, $\tilde{C} = CV$ and $\tilde{D} = D$.

The main idea is to split system (2.2.6) into the slow subsystem (2.2.7) and the fast subsystem (2.2.8) as discussed in Section 2.2.2 and reduce them separately. Considering the additive decomposition (2.2.9) of the transfer function $\mathbf{G}$ of (2.2.6), this is equivalent to a separate approximation of the strictly proper part $\mathbf{G}_{sp}$ and the polynomial part $\mathbf{G}_p$ by $\tilde{\mathbf{G}}_{sp}$ and $\tilde{\mathbf{G}}_p$, respectively. The resulting approximation is then given by

$$\tilde{\mathbf{G}} = \tilde{\mathbf{G}}_{sp} + \tilde{\mathbf{G}}_p.$$

Using the proper and improper Gramians, we now introduce the proper and improper Hankel singular values. Let n_f and n_∞ are the dimensions of the deflating subspaces of the pencil $\lambda E - A$ corresponding to the finite eigenvalues and the eigenvalue at infinity. Then the *proper Hankel singular values* σ_j are the square roots of the n_f largest eigenvalues of $G_{pc}E^T G_{po}E$, i.e.,

$$\sigma_j = \sqrt{\lambda_j(G_{pc}E^T G_{po}E)}.$$

The *improper Hankel singular values* θ_j are the square roots of the n_∞ largest eigenvalues of $G_{ic}A^T G_{io}A$, i.e.,

$$\theta_j = \sqrt{\lambda_j(G_{ic}A^T G_{io}A)}.$$

Note that if $E = I$, then the proper Hankel singular values are just Hankel singular values.

System (2.2.6) is called *balanced* if

$$G_{pc} = G_{po} = \begin{bmatrix} \Sigma & 0 \\ 0 & 0 \end{bmatrix} \qquad \text{and} \qquad G_{ic} = G_{io} = \begin{bmatrix} 0 & 0 \\ 0 & \Theta \end{bmatrix}$$

with $\Sigma = \operatorname{diag}(\sigma_1, \ldots, \sigma_{n_f})$ and $\Theta = \operatorname{diag}(\theta_1, \ldots, \theta_{n_\infty})$. As in BT for standard state space systems, we transform (2.2.6) into a balanced form and truncate those states which correspond to small proper Hankel values. This implies the reduction of the slow subsystem. Unfortunately, the reduction of the fast subsystem by truncation of the states corresponding to small nonzero improper Hankel singular values may lead to large errors. In the time domain, this corresponds to an approximation of the (hidden) constraints and may result in physically meaningless system. Examples can be found in [LS01, Sty11]. This difficulty can be overcame by finding a minimal realization of $\mathbf{G}_p$ as suggested in [Sty04].

The BT method for descriptor systems is summarized in Algorithm 2.2.

Similarly to the standard state space case, we can prove the following properties for the resulting reduced-order descriptor system.

Theorem 2.21. *Let a descriptor system* (2.2.6) *be asymptotically stable and let* $\mathbf{G}(s) = C(sE - A)^{-1}B + D$ *be its transfer function. Furthermore, let a reduced-order model* (2.4.5) *be computed by Algorithm 2.2. Then*

1. *the reduced system* (2.4.5) *is asymptotically stable and balanced;*

2. *for the transfer function* $\tilde{\mathbf{G}}(s) = \tilde{C}(s\tilde{E} - \tilde{A})^{-1}\tilde{B} + \tilde{D}$ *of* (2.4.5), *we have the error bound*

$$\|\mathbf{G} - \tilde{\mathbf{G}}\|_{\mathcal{H}_\infty} \leqslant 2(\sigma_{\eta+1} + \ldots + \sigma_{n_f});$$

3. *the index of the reduced system* (2.4.5) *does not exceed the index of the original system* (2.2.6).

Proof. See [Sty04]. □

Algorithm 2.2: Balanced truncation for descriptor systems

Input : a asymptotically stable system (E, A, B, C, D)

Output: a reduced-order asymptotically stable system $(\tilde{E}, \tilde{A}, \tilde{B}, \tilde{C}, \tilde{D})$

1 Solve the projected continuous-time Lyapunov equations

$$EG_{pc}A^T + AG_{pc}E^T = -P_l BB^T P_l^T, \qquad G_{pc} = P_r G_{pc} P_r^T, \tag{2.4.6}$$

$$E^T G_{po} A + A^T G_{po} E = -P_r^T C^T C P_r, \qquad G_{po} = P_l^T G_{po} P_l, \tag{2.4.7}$$

for the Cholesky factorizations $G_{po} = Z_{po} Z_{po}^T$ and $G_{pc} = Z_{pc} Z_{pc}^T$.

2 Solve the projected discrete-time Lyapunov equations

$$AG_{ic}A^T - EG_{ic}E^T = Q_l BB^T Q_l^T, \qquad G_{ic} = Q_r G_{ic} Q_r^T,$$

$$A^T G_{io} A - E^T G_{io} E = Q_r^T C^T C Q_r, \qquad G_{io} = Q_l^T G_{io} Q_l,$$

for the Cholesky factorizations $G_{io} = Z_{io} Z_{io}^T$ and $G_{ic} = Z_{ic} Z_{ic}^T$.

3 Compute the SVD

$$Z_{po}^T E Z_{pc} = \begin{bmatrix} U_1 & U_0 \end{bmatrix} \begin{bmatrix} \Sigma_1 & \\ & \Sigma_0 \end{bmatrix} \begin{bmatrix} V_1 & V_0 \end{bmatrix}^T,$$

where $\left[U_1, U_0\right]$ and $\left[V_1, V_0\right]$ are orthogonal, $\Sigma_1 = \mathrm{diag}(\sigma_1, \ldots, \sigma_{\eta_f})$, $\Sigma_0 = \mathrm{diag}(\sigma_{\eta_f}, \ldots, \sigma_{n_f})$.

4 Compute the SVD

$$Z_{io}^T A Z_{ic} = U_3 \Theta V_3^T,$$

where U_3 and V_3 have orthogonal columns and $\Theta = \mathrm{diag}(\theta_1, \ldots, \theta_{\eta_\infty})$ is nonsingular.

5 Compute the projection matrices

$$W = \left[Z_{po} U_1 \Sigma_1^{-\frac{1}{2}} \quad Z_{io} U_3 \Theta^{-\frac{1}{2}} \right], \qquad V = \left[Z_{pc} V_1 \Sigma_1^{-\frac{1}{2}} \quad Z_{ic} V_3 \Theta^{-\frac{1}{2}} \right].$$

6 Compute the reduced matrices $\tilde{E} = W^T E V$, $\tilde{A} = W^T A V$, $\tilde{B} = W^T B$, $\tilde{C} = CV$ and $\tilde{D} = D$.

The most computationally expensive part in Algorithm 2.2 is the numerical solution of the projected Lyapunov equations. These equations can be solved using the generalized LR-ADI method and the generalized Smith method as presented in [Sty08]. For solving the projected continuous-time Lyapunov equations (2.4.6) and (2.4.7), one can also use rational Krylov subspace methods [SS12]. In all these methods, the projectors P_l, P_r and Q_l, Q_r are required in explicite form that restricts the application of balanced truncation to general descriptor systems. For specially structured problems, some modifications have been presented to overcome this difficulty in [BS16, FRM08, HSS08, SV18, USKB12]. They are all based on an implicit index reduction and on an equivalence between the Schur complement linear systems and the original system matrices.

2.4.2 Proper orthogonal decomposition

The most popular model reduction method for nonlinear systems is proper orthogonal decomposition (POD). This method was first introduced in [Sir87] and then used in many different application fields such as signal analysis and pattern recognition [Fuk90], optimal control [AH00, KV02, LT01, SK98], fluid dynamics and coherent structures [AHLS88, IR98, RF94] and inverse problems [BJWW00].

We consider a nonlinear system

$$\dot{x} = Ax + f(x), \qquad x(t_0) = x_0 \tag{2.4.8}$$

with given $A \in \mathbb{R}^{n\times n}$, nonlinear function $f : \mathbb{R}^n \to \mathbb{R}^n$ and an initial state $x_0 \in \mathbb{R}^n$. The main idea of POD is to generate samples of the trajectory $\{x(t_1), \ldots, x(t_{n_s})\}$, called snapshots, and find a best approximation to the solution in a η-dimensional subspace of the space $\operatorname{span}\{x(t_1), \ldots, x(t_{n_s})\} \subset \mathbb{R}^n$. In other words, we are searching for an orthonormal basis $\{\psi_i\}_{i=1}^{\eta}$ that solves the minimization problem

$$\min_{\{\psi_k\}_{i=1}^{\eta}} \sum_{j=1}^{n_s} \left\| x(t_j) - \sum_{i=1}^{\eta} (x^T(t_j)\psi_i)\psi_i \right\|^2 . \tag{2.4.9}$$

This problem can be solved by computing the SVD of the snapshot matrix $\mathcal{X} = \left[x(t_1), \ldots, x(t_{n_s})\right]$, see, e.g. [Vol13]. Let

$$\mathcal{X} = \begin{bmatrix} U_1 & U_0 \end{bmatrix} \begin{bmatrix} \Sigma_1 & \\ & \Sigma_0 \end{bmatrix} \begin{bmatrix} V_1 & V_0 \end{bmatrix}^T$$

be the SVD with $\Sigma_1 = \operatorname{diag}(\sigma_1, \ldots, \sigma_\eta)$, $\Sigma_0 = \operatorname{diag}(\sigma_{\eta+1}, \ldots, \sigma_{n_s})$ and

$$\sigma_1 \geqslant \ldots \geqslant \sigma_\eta > \sigma_{\eta+1} \geqslant \ldots \geqslant \sigma_{n_s}.$$

Then the columns of $U_1 \in \mathbb{R}^{n\times\eta}$ provide the solution to the minimization problem (2.4.9).

Assume that the solution x of system (2.4.8) lies approximately in the span of the snapshots. Then we approximate $x \approx U_1\tilde{x}$ and project the nonlinear system (2.4.8) with U_1. The resulting reduced system of dimension η takes the form

$$\dot{\tilde{x}} = U_1^T A U_1 \tilde{x} + U_1^T f(U_1\tilde{x}). \tag{2.4.10}$$

The reduced dimension η can be selected using a tolerance *tol* by taking the smallest number η that satisfies

$$\frac{\sigma_\eta}{\sigma_1} < tol. \tag{2.4.11}$$

For a detailed description of how POD is applied to finite-dimensional dynamic systems and partial differential equations, we refer to [Vol13]. A modified POD with a weighted inner product and an error analysis are also presented here. Furthermore, a brief explanation of POD and its application to optimal control problems can be found in [KV02].

2.4.3 Discrete empirical interpolation method

It should be noted that while the reduced matrix $U_1^T A U_1$ in (2.4.10) can be precomputed and stored, the computation of the reduced nonlinear function $U_1^T f(U_1\tilde{x})$ requires the evaluation of f for large dimensional vectors $U_1\tilde{x}$ and projection afterward at every time step. This is very inefficient and we do not gain a speedup as it was for linear time-invariant systems. In order to speed up the simulation of the reduced-order nonlinear system (2.4.10), we employ the discrete empirical interpolation method (DEIM) proposed first in [CS10]. This method allows us to compute the approximation to the function $U_1^T f$ by only evaluating some entities of f.

Our goal is now to approximate the nonlinearity

$$f(U_1\tilde{x}) \approx U_f c(U_1\tilde{x})$$

in a low-dimensional subspace spanned by the columns of $U_f \in \mathbb{R}^{n\times\kappa}$. The basis matrix U_f is constructed by applying POD to the snapshots $\{f(x(t_1), \ldots, f(x(t_{n_s})\}$ of the function f. Let $\mathcal{X}_f = [f(x(t_1)), \ldots, f(x(t_{n_s})]$ be the snapshot matrix. Then the basis matrix U_f can be determined from the SVD

$$\mathcal{X}_f = \begin{bmatrix} U_f & \hat{U}_f \end{bmatrix} \begin{bmatrix} \Sigma_f & 0 \\ 0 & \hat{\Sigma}_f \end{bmatrix} \begin{bmatrix} V_f & \hat{V}_f \end{bmatrix}^T,$$

where Σ_f contains κ dominant singular values of $\mathcal{X}_f$. In order to determine the coefficient vector $c(U_1\tilde{x})$, we define a selector matrix

$$\mathcal{S}_\mathcal{K} = \begin{bmatrix} e_{p_1}, \ldots, e_{p_\kappa} \end{bmatrix},$$

where e_i is the i-th column of the identity matrix and $\mathcal{K} = \{p_1, \ldots, p_\kappa\}$ is a selected index set with pairwise different $1 \leqslant p_i \leqslant n$. With this matrix we can find $c(U_1\tilde{x})$ such that

$$\mathcal{S}_\mathcal{K}^T f(U_1\tilde{x}) = (\mathcal{S}_\mathcal{K}^T U_f) c(U_1\tilde{x})$$

holds. This implies that the selected components of the approximation $U_f c(U_1\tilde{x})$ coincide with those of the function $f(U_1\tilde{x})$.

There are different ways to compute the selector matrix $\mathcal{S}_\mathcal{K}$ or, respectively, the selected index set $\mathcal{K}$. We use the greedy procedure as presented in Algorithm 2.3.

An alternative approach for computing the selector matrix $\mathcal{S}_\mathcal{K}$ was introduced in [DG16]. It is based on a QR decomposition of U_f^T with column pivoting

$$U_f^T \begin{bmatrix} \mathcal{S}_\mathcal{K} & \Pi \end{bmatrix} = QR,$$

Algorithm 2.3: DEIM greedy procedure

Input : $\{u_k\}_{k=1}^{\kappa} \subset \mathbb{R}^n$ linear independent
Output: an index set $\mathcal{K}$ and a selector matrix $\mathcal{S}_{\mathcal{K}}$

Find $p_1 = \arg \max\limits_{1 \leqslant i \leqslant n} |(u_1)_i|$.
Set $U_f = \big[u_1\big]$, $\mathcal{K} = \{p_1\}$ and $\mathcal{S}_{\mathcal{K}} = e_{p_1}$.
for $k = 2$ *to* κ **do**
  Solve $(\mathcal{S}_{\mathcal{K}}^T U_f)c = \mathcal{S}_{\mathcal{K}}^T u_k$ for c.
  Compute $r = u_k - U_f c$.
  Find $p_k = \arg \max\limits_{1 \leqslant i \leqslant n} \{|r_i|\}$.
  $U_f \leftarrow \big[U_f, u_k\big]$, $\mathcal{K} \leftarrow \mathcal{K} \cup \{p_k\}$, $\mathcal{S}_{\mathcal{K}} \leftarrow \big[\mathcal{S}_{\mathcal{K}}, e_{p_k}\big]$.
end

where $\big[\mathcal{S}_{\mathcal{K}}, \Pi\big]$ is a permutation matrix, Q is orthogonal and R is an upper triangular matrix.

Replacing f in (2.4.10) with its DEIM approximation $U_f(\mathcal{S}_{\mathcal{K}}^T U_f)^{-1}\mathcal{S}_{\mathcal{K}}^T f$, we obtain the reduced-order system

$$\dot{\hat{x}} = U_1^T A U_1 \hat{x} + W \hat{f}(\hat{x}), \qquad \hat{x}(t_0) = U_1^T x_0, \tag{2.4.12}$$

where

$$W = U_1^T U_f (\mathcal{S}_{\mathcal{K}}^T U_f)^{-1}, \quad \hat{f}(\hat{x}) = \mathcal{S}_{\mathcal{K}}^T f(U_1 \hat{x}).$$

Note that the time-independent matrices $U_1^T A U_1 \in \mathbb{R}^{\eta \times \eta}$ and $W \in \mathbb{R}^{\eta \times \kappa}$ can be precomputed and stored in the offline stage. Then in the online stage, we have to evaluate only κ components of the function $f(U_1\hat{x})$. If f depends only on a few components of $U_1\hat{x}$, then the computation cost for (2.4.12) is independent of the original dimension n. For integrating the nonlinear system (2.4.12) in time, we use a one-step or multistep method [HNW93] which involves the solution of a sequence of systems of nonlinear equations. For this purpose, we employ the Newton iteration which requires the Jacobi matrix $J_{\hat{f}}(\hat{x})$ of the nonlinear function $\hat{f}$ at $\hat{x}$. This matrix has the form

$$J_{\hat{f}}(\hat{x}) = \mathcal{S}_{\mathcal{K}}^T J_f(U_1\hat{x}) U_1,$$

where $J_f(U_1\hat{x})$ is the Jacobi matrix of f at $U_1\hat{x}$. For efficient evaluation of $J_{\hat{f}}(\hat{x})$, we can use the matrix discrete empirical interpolation method (MDEIM) as presented in [SW13, Wil16]. Let $\mathcal{S}_{\mathcal{K}}^T J_f : \mathbb{R}^n \to \mathbb{R}^{\kappa \times n}$ be a nonlinear matrix-valued function. The goal is to find an approximation

$$\mathcal{S}_{\mathcal{K}}^T J_f(U_1\hat{x}) \approx \sum_{k=1}^{\rho} V_k g_k(\hat{x}),$$

where $V_k \in \mathbb{R}^{\kappa \times n}$ are constant matrices, $g_k : \mathbb{R}^{\eta} \to \mathbb{R}$ continuous functions and ρ is small compared to n. Such an approximation can be determined by DEIM using the vectorization operator [WSH14]. Here, we use an efficient formulation of MDEIM from [Wil16]. The main advantage of this approach is that it avoids the vectorization. First, we collect the snapshots

$$J_1 = \mathcal{S}_{\mathcal{K}}^T J_f(x(t_1)), \ldots, J_{n_s} = \mathcal{S}_{\mathcal{K}}^T J_f(x(t_{n_s}))$$

and, following [Vol99], construct the matrix

$$\mathcal{X}_J = \begin{bmatrix} \langle J_1, J_1\rangle_F & \dots & \langle J_1, J_{n_s}\rangle_F \\ \vdots & \ddots & \vdots \\ \langle J_{n_s}, J_1\rangle_F & \dots & \langle J_{n_s}, J_{n_s}\rangle_F \end{bmatrix},$$

where $\langle J_i, J_j\rangle_F = \text{trace}(J_i^T J_j)$ denotes the Frobenius inner product of the matrices J_i and J_j. One can show that this matrix $\mathcal{X}_J$ is symmetric and positive semidefinite. Computing the eigenvalue decomposition (EVD)

$$\mathcal{X}_J = \begin{bmatrix} U_J & \hat{U}_J \end{bmatrix} \begin{bmatrix} \Lambda_1 & \\ & \Lambda_0 \end{bmatrix} \begin{bmatrix} U_J & \hat{U}_J \end{bmatrix}^T, \tag{2.4.13}$$

where $\Lambda_1 \in \mathbb{R}^{\rho\times\rho}$ contains ρ dominant eigenvalues of $\mathcal{X}_J$, the POD basis can then be determined by

$$V_j = \sum_{i=1}^{n_s} J_i u_{i,j}, \quad j = 1, \dots, \rho, \tag{2.4.14}$$

where $u_{i,j}$ are the entries of $U_J\Lambda_1^{-\frac{1}{2}} \in \mathbb{R}^{n_s\times\rho}$. By definition, the basis matrices V_j have the same sparsity pattern as $\mathcal{S}_\mathcal{K}^T J_f$. The following lemma shows that $V_1, \dots, V_\rho$ form an orthonormal system.

Lemma 2.22. *The matrices V_j in (2.4.14) fulfill $\langle V_l, V_j\rangle_F = \delta_{l,j}$, where $\delta_{l,j}$ is the Kronecker delta.*

Proof. Computing $\langle V_l, V_j\rangle_F$ by using (2.4.14) and the definition of the Frobenius scalar product, we obtain

$$\begin{aligned} \langle V_l, V_j\rangle_F &= \text{trace}\left(\left(\sum_{i=1}^{n_s} J_i u_{i,l}\right)^T \left(\sum_{k=1}^{n_s} J_k u_{k,j}\right)\right) \\ &= \sum_{i=1}^{n_s}\sum_{k=1}^{n_s} u_{i,l}\,\text{trace}\left(J_i^T J_k\right) u_{k,j} \\ &= \sum_{i=1}^{n_s}\sum_{k=1}^{n_s} u_{i,l}\langle J_i, J_k\rangle_F u_{k,j} \\ &= \left(\Lambda_1^{-\frac{1}{2}} U_J^T \mathcal{X}_J U_J \Lambda_1^{-\frac{1}{2}}\right)_{l,j} = \delta_{l,j}. \end{aligned}$$

□

Imposing the condition that the entries of the approximation coincide with those of $\mathcal{S}_\mathcal{K}^T J_f$ for a selected set of pairwise different indices

$$\mathcal{J} = \{(i_l, j_l) : 1 \leqslant i_l \leqslant \kappa,\ 1 \leqslant j_l \leqslant n \text{ for } l = 1, \dots, \rho\},$$

we get the equations

$$\left(\mathcal{S}_\mathcal{K}^T J_f(U_1\hat{x})\right)_{i_l,j_l} = \sum_{k=1}^{\rho} (V_k)_{i_l,j_l} g_k(\hat{x}), \quad (i_l, j_l) \in \mathcal{J}.$$

Algorithm 2.4: MDEIM greedy procedure

Input : $\{V_k\}_{k=1}^{\rho} \subset \mathbb{R}^{\kappa \times n}$ linear independent
Output: an index set $\mathcal{J} = \{(i_1, j_1), \ldots, (i_\rho, j_\rho)\}$, $G_\rho \in \mathbb{R}^{\rho\times\rho}$

1 Find $(i_1, j_1) = \underset{1\leqslant i\leqslant\kappa, 1\leqslant j\leqslant n}{\arg\max} |(V_1)_{i,j}|$.
2 Set $\mathcal{J} = \{(i_1, j_1)\}, G_1 = (V_1)_{i_1,j_1}$.
3 **for** $k = 2$ **to** ρ **do**
4 Set $b = \left[(V_k)_{i_1,j_1}, \ldots, (V_k)_{i_{k-1},j_{k-1}}\right]^T$.
5 Solve $G_{k-1}c = b$ for $c = \left[c_1, \ldots, c_{k-1}\right]^T$.
6 Compute $R_k = V_k - \sum_{l=1}^{k-1} V_l c_l$.
7 Find $(i_k, j_k) = \underset{1\leqslant i\leqslant\kappa, 1\leqslant j\leqslant n}{\arg\max} |(R_k)_{i,j}|$.
8 Set $\mathcal{J} \leftarrow \mathcal{J} \cup \{(i_k, j_k)\}$.
9 Set $G_k = \begin{bmatrix} G_{k-1} & b \\ v & (V_k)_{i_k,j_k}\end{bmatrix}$ with $v = \left[(V_1)_{i_k,j_k}, \ldots, (V_{k-1})_{i_k,j_k}\right]$.
10 **end**

These equations can shortly be written as a linear system

$$G_\rho \begin{bmatrix} g_1(\hat{x}) \\ \vdots \\ g_\rho(\hat{x}) \end{bmatrix} = \begin{bmatrix} \left(\mathcal{S}_\mathcal{K}^T J_f(U_1\hat{x})\right)_{i_1,j_1} \\ \vdots \\ \left(\mathcal{S}_\mathcal{K}^T J_f(U_1\hat{x})\right)_{i_\rho,j_\rho} \end{bmatrix}$$

with

$$G_\rho = \begin{bmatrix} (V_1)_{i_1,j_1} & \cdots & (V_\rho)_{i_1,j_1} \\ \vdots & \ddots & \vdots \\ (V_1)_{i_\rho,j_\rho} & \cdots & (V_\rho)_{i_\rho,j_\rho} \end{bmatrix}.$$

If G_ρ is nonsingular, then the functions $g_k(\hat{x})$ can be calculated as

$$\begin{bmatrix} g_1(\hat{x}) \\ \vdots \\ g_\rho(\hat{x}) \end{bmatrix} = G_\rho^{-1} \begin{bmatrix} \left(\mathcal{S}_\mathcal{K}^T J_f(U_1\hat{x})\right)_{i_1,j_1} \\ \vdots \\ \left(\mathcal{S}_\mathcal{K}^T J_f(U_1\hat{x})\right)_{i_\rho,j_\rho} \end{bmatrix}.$$

The index set $\mathcal{J}$ is determined using the greedy procedure as presented in Algorithm 2.4 which is a generalization of Algorithm 2.3 to the matrix case.

The following lemma establishes that the matrices G_k in Algorithm 2.4 are nonsingular implying that Step 5 is well defined.

Lemma 2.23. *The matrices G_k, $k = 1, \ldots, \rho$, determined in Algorithm 2.4 are nonsingular.*

Proof. The result can be proven by induction. The matrix $G_1 = (V_1)_{i_1,j_1}$ is nonsingular, since $V_1 \neq 0$. Assume that G_{k-1} is nonsingular for $k > 1$. Then $G_{k-1}c = b$ gives

$$G_k \begin{bmatrix} I & -c \\ 0 & 1 \end{bmatrix} = \begin{bmatrix} G_{k-1} & 0 \\ v & (V_k)_{i_k,j_k} - vc \end{bmatrix}. \tag{2.4.15}$$

By Lemma 2.22 we have that

$$R_k = V_k - \sum_{l=1}^{k-1} V_l c_l \neq 0$$

and, therefore, $(R_k)_{i_k,j_k} = (V_k)_{i_k,j_k} - vc \neq 0$. The matrix on the right-hand side of equation (2.4.15) is a lower triangular matrix with invertible diagonal matrices and, therefore, nonsingular. This means that G_k is nonsingular. □

Next, we present error estimates for the DEIM approximation and for the POD-DEIM reduced model (2.4.12).

Theorem 2.24. *Let $f : \mathcal{D} \to \mathbb{R}^n$ be a nonlinear vector-valued function with a domain $\mathcal{D} \subset \mathbb{R}^n$ and let $U_f(\mathcal{S}_\mathcal{K}^T U_f)^{-1}\mathcal{S}_\mathcal{K}^T f$ be a DEIM approximation of f, where $U_f := [u_1, \ldots, u_\kappa]$ is a DEIM basis with $\kappa \in \{1, \ldots, n_s\}$ and $\mathcal{S}_\mathcal{K}$ is a selector matrix as in Algorithm 2.3. Additionally, let $x \in \mathcal{D}$ be arbitrary such that $f(x)$ is in the space spanned by the columns of U_f. Then an error bound for the DEIM approximation is given by*

$$\|f(x) - U_f(\mathcal{S}_\mathcal{K}^T U_f)^{-1}\mathcal{S}_\mathcal{K}^T f(x)\| \leqslant \|(\mathcal{S}_\mathcal{K}^T U_f)^{-1}\|_2 \mathcal{E}_*(f(x))$$

with $\mathcal{E}_*(f(x)) = \|(I - U_f U_f^T) f(x)\|_2$.

Proof. See [CS10, Lemma 3.2]. □

Since $\mathcal{E}_*(f(x))$ depends on $x \in \mathcal{D}$, it is hard to compute. Therefore, it was proposed in [CS10] to approximate

$$\mathcal{E}_* = \mathcal{E}_*(f(x)) \lesssim \sum_{i=\kappa+1}^{n_s} \sigma_i(\mathcal{X}_f),$$

where $\sigma_i(\mathcal{X}_f)$ are the truncated singular values of the snapshot matrix $\mathcal{X}_f$. This approximation is reasonable as long as $f(x)$ is nearly in the range of $\mathcal{X}_f$.

In order to estimate the error $x - \hat{x}$, where x and $\hat{x}$ are the solutions of the nonlinear system (2.4.8) and the POD-DEIM reduced system (2.4.12), respectively, we need a concept of logarithmic Lipschitz constants for linear and nonlinear functions. Let $G \in \mathbb{R}^{n \times n}$ be a symmetric and positive definite matrix. Then $\langle x, z \rangle_G = z^T G x$ defines a scalar product on $\mathbb{R}^n$. The corresponding vector norm is given by $\|x\|_G = \sqrt{x^T G x}$ for $x \in \mathbb{R}^n$ and the induced matrix norm $\|A\|_G = \sup_{\|x\|_G = 1} \|Ax\|_G$ for $A \in \mathbb{R}^{n \times n}$. For a Lipschitz continuous function $f : \mathbb{R}^n \times \mathbb{R}^n$, the *logarithmic Lipschitz constant with respect to G* is defined as

$$L_G[f] = \sup_{\substack{x,z \in \mathbb{R}^n \\ x \neq z}} \frac{\langle x - z, f(x) - f(z) \rangle_G}{\|x - z\|_G^2}.$$

Furthermore, the *local logarithmic Lipschitz constant* for f at $z \in \mathbb{R}^n$ with respect to G is given by

$$L_G[f](z) = \sup_{x \in \mathbb{R}^n \setminus \{z\}} \frac{\langle x - z, f(x) - f(z) \rangle_G}{\|x - z\|_G^2}. \tag{2.4.16}$$

For $G = I$, $L_I[f]$ is also denoted by $L_2[f]$. For a linear function $f_A(x) = Ax$ with $A \in \mathbb{R}^{n\times n}$, the logarithmic Lipschitz constant takes the form

$$L_G[A] = \sup_{\substack{x,z\in\mathbb{R}^n \\ x\neq z}} \frac{\langle x-z, A(x-z)\rangle_G}{\|x-z\|_G^2} = \sup_{x\in\mathbb{R}^n\setminus\{0\}} \frac{\langle x, Ax\rangle_G}{\|x\|_G^2}.$$

For $G = I$, we obtain

$$L_I[A] = \sup_{x\in\mathbb{R}^n\setminus\{0\}} \frac{\langle x, Ax\rangle}{\|x\|^2} = \lambda_{\max}\left(\frac{A+A^T}{2}\right) =: L_2[A], \tag{2.4.17}$$

where $\lambda_{\max}(\cdot)$ denotes the largest eigenvalue of the corresponding matrix. The value $L_2[A]$ is known in the literature as *logarithmic norm* [Dah59, Söd06]. It is not a norm in the usual sense, since, for example, $L_2[A] < 0$ for a symmetric negative definite matrix A.

The following theorem gives an a priori error estimate for $x - \hat{x}$.

Theorem 2.25. *Let x be the solution of the full-order ODE system* (2.4.8) *and $\hat{x}$ be the solution of the POD-DEIM reduced system* (2.4.12) *on the time interval* $[0,T]$. *Let U_1 be a POD basis and $W = U_1^T U_f (\mathcal{S}_\mathcal{K}^T U_f)^{-1}$, where U_f and $\mathcal{S}_\mathcal{K}$ are the DEIM basis and selector matrices. Assume that the logarithmic norm of A defined as in* (2.4.17) *satisfies $L_2[A] < 0$ and that f is Lipschitz continuous with Lipschitz constant L_f. Then*

$$\int_0^T \|x(t) - U_1\hat{x}(t)\|^2\,dt \leqslant \max\{1 + c\alpha^2 T, c\beta^2 T\}(\mathcal{E}_x + \mathcal{E}_f),$$

where

$$\begin{aligned}
\alpha &= \|U_1^T A\|_2 + L_f\|W\mathcal{S}_\mathcal{K}^T\|_2,\\
\beta &= \|U_1^T - W\mathcal{S}_\mathcal{K}^T\|_2,\\
\gamma &= L_f\|W\mathcal{S}_\mathcal{K}^T\|_2,\\
c &= \frac{e^{2\gamma(e^{L_2[A]}-1)/L_2[A]}}{|L_2[A]|},\\
\mathcal{E}_x &= \int_0^T \|x(t) - U_1U_1^T x(t)\|^2\,dt = \sum_{i=\eta+1}^{n_s} \sigma_i(\mathcal{X}),\\
\mathcal{E}_f &= \int_0^T \|f(x(t)) - U_fU_f^T f(x(t))\|^2\,dt = \sum_{i=\kappa+1}^{n_s} \sigma_i(\mathcal{X}_f),
\end{aligned}$$

with $\sigma_i(\mathcal{X})$ and $\sigma_i(\mathcal{X}_f)$ denoting the truncated singular values of the snapshot matrices $\mathcal{X}$ and $\mathcal{X}_f$, respectively.

Proof. See [CS12, Theorem 3.1]. □

Instead of an *a priori error estimate*, as in Theorem 2.25, an *a posteriori error estimate* may be used as presented in the following theorem.

Theorem 2.26. *Let a DEIM basis $\{u_1, \ldots, u_m\}$ and a DEIM index set $\{p_1, \ldots, p_m\}$ be given. Assume that the m-order DEIM approximation of f is exact. For $\kappa \leqslant m-1$ and $\hat{\kappa} = m - \kappa$, set*

$$\mathcal{S}_{\mathcal{K}} = [e_{p_1}, \ldots, e_{p_\kappa}], \qquad \hat{\mathcal{S}}_{\mathcal{K}} = [e_{p_{\kappa+1}}, \ldots, e_{p_{\kappa+\hat{\kappa}}}],$$
$$U_f = [u_1, \ldots, u_\kappa], \qquad \hat{U}_f = [u_{\kappa+1}, \ldots, u_{\kappa+\hat{\kappa}}].$$

Let x be the solution of the full-order ODE system (2.4.8) *and let $\hat{x}$ be the solution of the POD-DEIM reduced system* (2.4.12) *with a POD basis U_1 and the DEIM basis and selector matrices U_f and $\mathcal{S}_{\mathcal{K}}$, respectively. Then the state error $x - U_1\hat{x}$ is bounded as*

$$\|x(t) - U_1\hat{x}(t)\|_G \leqslant \int_0^t \beta(\tau) e^{\int_\tau^t \alpha(\sigma)\, d\sigma}\, d\tau + e^{\int_0^t \alpha(\sigma)\, d\sigma} \|x(0) - U_1 U_1^T x(0)\|_G$$

for $t \in [0, T]$, where

$$\begin{aligned}
\alpha(t) &= L_G[F](U_1\hat{x}(t)), \quad F(x) = Ax + f(x),\\
\beta(t) &= \|M_1 \mathcal{S}_{\mathcal{K}}^T f(U_1\hat{x}(t)) - M_2 \hat{\mathcal{S}}_{\mathcal{K}}^T f(U_1\hat{x}(t)) + (I - U_1U_1^T)AU_1\hat{x}\|_G,\\
M_2 &= \left(U_f \left(\mathcal{S}_{\mathcal{K}}^T U_f\right)^{-1} \mathcal{S}_{\mathcal{K}}^T \hat{U}_f - \hat{U}_f\right) \left(\hat{\mathcal{S}}_{\mathcal{K}}^T \hat{U}_f - \hat{\mathcal{S}}_{\mathcal{K}}^T U_f \left(\mathcal{S}_{\mathcal{K}}^T U_f\right)^{-1} \mathcal{S}_{\mathcal{K}}^T \hat{U}_f\right)^{-1},\\
M_1 &= M_2 \hat{\mathcal{S}}_{\mathcal{K}}^T U_f \left(\mathcal{S}_{\mathcal{K}}^T U_f\right)^{-1} + (I - U_1U_1^T)U_f \left(\mathcal{S}_{\mathcal{K}}^T U_f\right)^{-1}.
\end{aligned}$$

Proof. See [WSH14]. □

The computation of the local logarithmic Lipschitz constant $L_G[F](U_1\hat{x})$ and the error estimator in Theorem 2.26 has been discussed in [WSH14]. It relies on an approximation of $L_G[F](U_1\hat{x}(t))$ with a logarithmic norm $L_G[J_F(U_1\hat{x}(t))]$ of the Jacobi matrix of F evaluated at $U_1\hat{x}(t)$ which can be calculated in an efficient way using the offline/online decomposition.

2.5 Finite element method

The finite element method (FEM) is a numerical technique for solving boundary value problems for partial differential equations (PDEs). In this section, we give an outline of this method using Poisson's equation as an example. For detailed description as well as convergence and error analysis for different types of PDEs, we refer to [Cia02, Eva98, QV94, Zei90a, Zei90b, Zul08].

First, we define some functional spaces and weak derivatives that will be used in the following.

Definition 2.27. Let $\Omega \in \mathbb{R}^d$ be a bounded open domain and $1 \leqslant p < \infty$. The functional space defined by

$$\mathcal{L}^p(\Omega) = \left\{ w : \Omega \to \mathbb{R}^m \text{ measurable} : \int_\Omega \|w(\xi)\|^p \, d\xi < \infty \right\}$$

with the norm

$$\|w\|_{\mathcal{L}^p(\Omega)} = \left(\int_\Omega \|w(\xi)\|^p \, d\xi \right)^{\frac{1}{p}}$$

is called *Lebesgue space.*

Definition 2.28. Let

$$\mathcal{L}^1_{\text{loc}}(\Omega) = \left\{ w : \Omega \to \mathbb{R}^m \,\middle|\, w \in \mathcal{L}^1(K) \text{ for all compact subsets } K \subset \Omega \right\}$$

be a space of locally integrable functions and let $\alpha = (\alpha_1, \ldots, \alpha_l) \in \mathbb{N}_0^l$ be a multi-index with $|\alpha| = \alpha_1 + \ldots + \alpha_l$. A function $v \in \mathcal{L}^1_{\text{loc}}(\Omega)$ is called the *α-th weak derivative* of $w \in \mathcal{L}^1_{\text{loc}}(\Omega)$ if

$$\int_\Omega w \, \mathcal{D}^\alpha \psi \, d\xi = (-1)^{|\alpha|} \int_\Omega v \psi \, d\xi$$

for all test functions $\psi \in \mathcal{C}_c^\infty(\Omega)$, where $\mathcal{C}_c^\infty(\Omega)$ is the space of infinitely often differentiable functions with compact support in Ω.

We now define the Sobolev spaces which play a fundamental role in FEM.

Definition 2.29. Let $1 \leqslant p < \infty$ and $k \in \mathbb{N}$. The *Sobolev space* $\mathcal{W}^{k,p}$ is given by

$$\mathcal{W}^{k,p}(\Omega) = \left\{ w \in \mathcal{L}^p(\Omega) : \quad \mathcal{D}^\alpha w \in \mathcal{L}^p(\Omega) \text{ for all } \alpha \in \mathbb{N}_0^k \right\}$$

with the norm

$$\|w\|_{\mathcal{W}^{k,p}(\Omega)} = \left(\sum_{|\alpha| \leqslant k} \|\mathcal{D}^\alpha w\|^p_{\mathcal{L}^p(\Omega)} \right)^{\frac{1}{p}}.$$

For $p = 2$, we use the notation

$$H^k(\Omega) = \mathcal{W}^{k,2}(\Omega).$$

Note that $H^k(\Omega)$ is a Hilbert space with respect to the scalar product

$$\langle w, v \rangle_{H^k(\Omega)} = \int_\Omega \sum_{|\alpha| \leqslant k} \mathcal{D}^\alpha w \, \mathcal{D}^\alpha v \, dx,$$

see [QV94, Section 1]. As the boundary $\partial\Omega$ of Ω is a null set, $v \in \mathcal{W}^{k,p}(\Omega)$ may be undefined or not uniquely determined on $\partial\Omega$. Therefore, we need a trace operator which makes it possible to define a unique restriction of v to the boundary.

Theorem 2.30 (Trace theorem). *Let Ω be a bounded open domain with $\mathcal{C}^1$-boundary $\partial\Omega$. Then there exists a bounded linear operator*

$$\tau : \mathcal{W}^{1,p}(\Omega) \to \mathcal{L}^p(\partial\Omega)$$

such that

1. $\tau w = w|_{\partial\Omega}$ *if* $w \in \mathcal{W}^{1,p}(\Omega) \cap \mathcal{C}^1(\bar{\Omega})$,
2. $\|\tau w\|_{\mathcal{L}^p(\Omega)} \leqslant \mathrm{C}\|w\|_{\mathcal{W}^{1,p}(\Omega)}$ *for all* $w \in \mathcal{W}^{1,p}(\Omega)$ *with a constant* $\mathrm{C} \geqslant 0$ *depending on* p *and* Ω.

Proof. See [Eva98, Chapter 5.5]. □

The operator τ in Theorem 2.30 is called the *trace operator.* Using this operator, we define the Sobolev spaces

$$\mathcal{W}_0^{k,p}(\Omega) = \{w \in \mathcal{W}^{k,p}(\Omega) \; : \; \tau w = 0 \text{ on } \partial\Omega\}$$

and $H_0^1(\Omega) = \mathcal{W}_0^{1,2}(\Omega)$.

2.5.1 Poisson's equation

Consider now Poisson's equation

$$-\triangle z = f \qquad \text{in } \Omega, \tag{2.5.1a}$$

with homogeneous Dirichlet and Neumann boundary conditions

$$z = 0 \qquad \text{on } \Gamma_1, \tag{2.5.1b}$$

$$\frac{\partial z}{\partial n_0} = 0 \qquad \text{on } \Gamma_2, \tag{2.5.1c}$$

where $\Omega \subset \mathbb{R}^d$ is a bounded connected open domain with a boundary $\partial\Omega = \Gamma_1 \cup \Gamma_2$, $\Gamma_1 \cap \Gamma_2 = \emptyset$, and n_0 is an outer normal to Γ_2. Often, a strong formulation as in (2.5.1) leads to unnecessarily strong conditions on the smoothness of the solution, which may not exist. To overcome this difficulty, a weak formulation of (2.5.1) is used. In order to derive the weak formulation, for the Poisson problem (2.5.1), we define the trial space

$$\mathcal{V} = \{\psi \in H^1(\Omega) : \tau\psi = 0 \text{ on } \Gamma_1\} \tag{2.5.2}$$

and the test space $\mathcal{D}(\Omega) = \mathcal{V}$.

Multiplying equation (2.5.1a) with $\psi \in \mathcal{D}(\Omega)$ and integrating over the domain Ω results in

$$\int_\Omega -\triangle z\, \psi \, d\xi = \int_\Omega f\psi \, d\xi.$$

Then using Green's formula (see [Eva98, Section C2]) this equation can be rewritten as

$$\int_\Omega \nabla z \cdot \nabla\psi \, d\xi - \int_{\partial\Omega} \frac{\partial z}{\partial n_0}\psi \, ds = \int_\Omega f\psi \, d\xi. \tag{2.5.3}$$

Taking into account that

$$\int_{\Gamma_2} \frac{\partial z}{\partial n_0} \psi \, ds = 0$$

due to the Neumann boundary condition (2.5.1c) and

$$\int_{\Gamma_1} \frac{\partial z}{\partial n_0} \psi \, ds = 0$$

for $\psi \in \mathcal{V}$, equation (2.5.3) is simplified to

$$\int_\Omega \nabla z \cdot \nabla \psi d\xi = \int_\Omega f \psi \, d\xi. \tag{2.5.4}$$

This equation is the weak formulation of (2.5.1). A function $z \in \mathcal{V}$ is called a *weak solution* of (2.5.4), if z satisfies (2.5.4) for all $\psi \in \mathcal{D}(\Omega)$. Introducing a bilinear form

$$b(z, \psi) = \int_\Omega \nabla z \cdot \nabla \psi \, d\xi \tag{2.5.5}$$

and a linear functional

$$l(\psi) = \int_\Omega f \psi \, d\xi,$$

equation (2.5.4) can shortly be written as

$$b(z, \psi) = l(\psi), \quad \psi \in \mathcal{D}(\Omega). \tag{2.5.6}$$

Definition 2.31. Let $\mathcal{H}$ be a Hilbert space. A bilinear form

$$b : \mathcal{H} \times \mathcal{H} \to \mathbb{R}$$

is called

1. *bounded* if there exists a constant $L > 0$ such that

$$|b(v, w)| \leqslant L \|v\|_\mathcal{H} \|w\|_\mathcal{H}$$

for all $v, w \in \mathcal{H}$,

2. *coercive* if there exists a constant $m > 0$ such that

$$b(v, v) \geqslant m \|v\|^2_\mathcal{H},$$

for all $v \in \mathcal{H}$.

The existence and uniqueness of the solution of equation (2.5.6) can be established using the Lax-Milgram lemma.

Theorem 2.32 (Lax-Milgram Lemma). *Let $\mathcal{H}$ be a (real) Hilbert space and let $\mathcal{H}'$ be its dual space. Consider a bilinear form $b : \mathcal{H} \times \mathcal{H} \to \mathbb{R}$ and a linear continuous functional $l : \mathcal{H} \to \mathbb{R}$. Assume that b is bounded and coercive. Then there exists an unique solution $v \in \mathcal{H}$ of*

$$b(v, \psi) = l(\psi) \qquad \textit{for all } \psi \in \mathcal{D}(\Omega),$$

with

$$\|v\|_{\mathcal{H}} \leqslant \frac{1}{m}\|l\|_{\mathcal{H}'}$$

where m is the coercivity constant.

Proof. See [QV94, Theorem 5.1.1]. □

To show the coercivity of the bilinear form b in (2.5.5), we need the following theorem.

Theorem 2.33 (Poincaré's inequality). *Let $\Omega \subset \mathbb{R}^d$ be a bounded, connected, open domain with a non-empty subset Γ_1 of the C^1-boundary $\partial\Omega$ and let $1 \leqslant p \leqslant \infty$. Then there exists a constant $C > 0$ depending only on d, p and Ω such that*

$$\|w\|_{\mathcal{L}^p(\Omega)} \leqslant C\|\nabla w\|_{\mathcal{L}^p(\Omega)}$$

for all $w \in \mathcal{W}^{1,p}(\Omega)$ with $\tau w = 0$ on Γ_1.

Proof. [QV94, Theorem 1.3.3] □

It follows from Theorem 2.33 that $\|w\|_{\mathcal{V}} \leqslant \tilde{C}\|\nabla w\|_{\mathcal{L}^2(\Omega)}$ for a constant $\tilde{C} = \sqrt{1 + C^2} > 0$. Then the bilinear form b fulfills

$$b(w, w) = \langle \nabla w, \nabla w \rangle_{\mathcal{L}^2(\Omega)} = \|\nabla w\|^2_{\mathcal{L}^2(\Omega)} \geqslant \tilde{C}^{-2}\|w\|^2_{\mathcal{V}} \quad \text{for all } w \in \mathcal{V}.$$

and, therefore, it is coercive. Moreover, using the Hölder inequality (see [QV94, Chapter 1]), we obtain

$$|b(v, w)| \leqslant \langle \nabla v, \nabla w \rangle_{\mathcal{L}^2(\Omega)} \leqslant \|v\|_{\mathcal{V}}\|w\|_{\mathcal{V}} \quad \text{for all } w \in \mathcal{V}.$$

This means that b is bounded. Thus, Theorem 2.33 implies the existence and uniqueness of the solution of Poisson's problem (2.5.1).

2.5.2 Discretization

As we cannot solve the problem (2.5.6) in the infinite-dimensional case numerically, our goal is now to find a finite-dimensional approximation to the test and trial spaces and to discretize equation (2.5.6). For this purpose, we first introduce a regular triangulation of the domain Ω.

Definition 2.34. Let $\Omega \subset \mathbb{R}^d$ be a bounded open domain with a polygonal boundary. A set $\mathcal{T} = \{K_1, \dots, K_{n_\mathcal{T}}\}$ is called *regular triangulation* of Ω, if all elements $K_i \in \mathcal{T}$ are closed simplicies in $\mathbb{R}^d$ with

$$\operatorname{int}(K_i) \cap \operatorname{int}(K_j) = \emptyset, \quad i \neq j,$$
$$\bigcup_{i=1}^{n_\mathcal{T}} K_i = \bar{\Omega},$$
$$K_i \cap K_j = \begin{cases} \emptyset & \text{or} \\ \text{a common full edge} & \text{or} \\ \text{a common vertex.} \end{cases}$$

Here $\operatorname{int}(K_i)$ denotes the interior of K_i.

In a second step, we define a finite element (K, P, Σ).

Definition 2.35. A *finite element* is defined by a triple (K, P, Σ), where

- K is a bounded, closed subset of $\mathbb{R}^d$ with nonempty interior and piecewise smooth boundary;
- $P = P(K)$ is a finite-dimensional function space on K of dimension n_P,
- $\Sigma = \{\varsigma_1, \dots, \varsigma_{n_P}\}$ is a basis for the dual space P'.

To obtain a finite-dimensional space $\mathcal{V}_h \subset \mathcal{V}$ from the set of finite elements $\{(K, P, \Sigma)\}_{K \in \mathcal{T}}$, we introduce a reference finite element $(\hat{K}, \hat{P}, \hat{\Sigma})$ and a set of affine mappings $F_K : \hat{K} \to K$ given by $F_K(\hat{\xi}) = B_K \hat{\xi} + b_K$ with an invertible matrix $B_K \in \mathbb{R}^{d \times d}$ and $b_K \in \mathbb{R}^d$. These mappings map the reference cell $\hat{K}$ to the cells K, i.e., $K = F_K(\hat{K})$ for all $K \in \mathcal{T}$. From the conditions $\hat{\varsigma}_k(\hat{\varphi}_l) = \delta_{k,l}$ for $k, l = 1, \dots, n_P$, the nodal basis functions $\hat{\varphi}_l$, $l = 1, \dots, n_P$, on $\hat{K}$ can be determined. Using these functions and the reference mapping F_K, we can then generate the nodal basis functions

$$\varphi_j = \hat{\varphi}_j \circ F_K^{-1}, \qquad j = 1, \dots, n_P,$$

defined on K and, finally, the global basis functions $\{\phi_j\}_{j=1}^n$ for the discrete function space $\mathcal{V}_h = \mathcal{V}_h(\mathcal{T})$ of dimension n. In order to discretize the variational problem (2.5.6) for Poisson's equation, we consider the Lagrange finite elements. The Lagrange reference finite element is given by a triple $(\hat{K}, \hat{P}, \hat{\Sigma})$, where $\hat{K}$ is a reference simplex in $\mathbb{R}^d$ with the vertices $v_1, \dots, v_{d+1} \in \mathbb{R}^d$, $\hat{P} = \mathbb{P}_q(\hat{K})$ is the space of polynomials of degree q on $\hat{K}$ and

$$\hat{\Sigma} = \{\hat{\varsigma}_k : \mathbb{P}_q \to \mathbb{R} : \hat{\varsigma}_k(v) = v(\zeta^k),\ k = 1, \dots, n_P, \text{ for all } v \in \hat{P}\}$$

where

$$\zeta^k \in \left\{ \zeta = \sum_{i=1}^{d+1} \mu_i v_i \in \mathbb{R}^d : \sum_{i=1}^{d+1} \mu_i = 1,\ \mu_i \in \left\{0, \frac{1}{q}, \dots, \frac{q-1}{q}, 1\right\} \right\}.$$

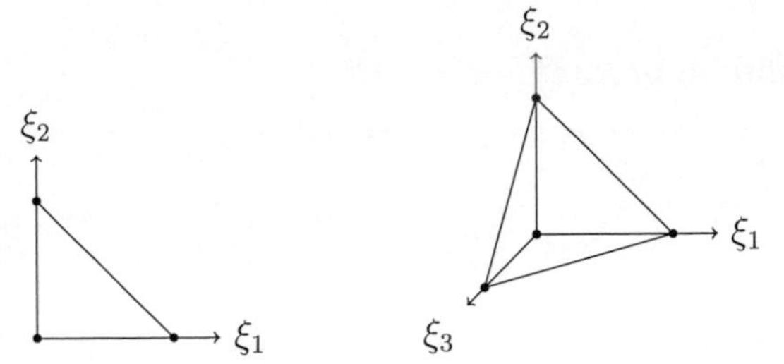

(a) Reference triangle in 2D

(b) Reference tetrahedron in 3D

Figure 2.1: Reference domain K

The reference simplex $\hat{K}$ is a triangle in $\mathbb{R}^2$ and a tetrahedron in $\mathbb{R}^3$ as shown in Figure 2.1(a) and 2.1(b), respectively. For $q = 1$ the points ζ^k are the vertices of $\hat{K}$ (they are marked by dot's in Figure 2.1) and for $q = 2$, these points are the vertices and edge midpoints of $\hat{K}$. Note that conditions $\hat{\varsigma}_k(\hat{\varphi}_l) = \hat{\varphi}_l(\zeta^k) = \delta_{k,l}$ define the nodal basis $\{\hat{\varphi}_l\}_{l=1}^{n_P}$ of $\hat{P}$ uniquely. It is given by $\{1 - \zeta_1, -\zeta_2, \zeta_1, \zeta_2\}$ in 2D and$\{1 - \zeta_1, -\zeta_2 - \zeta_3, \zeta_1, \zeta_2, \zeta_3\}$ in 3D.

Let

$$\mathcal{V}_h = \{v \in \mathcal{C}(\bar{\Omega}) : v|_K \in \mathbb{P}_1(K) \text{ for all } K \subset \mathcal{T}\} \subset H^1(\Omega)$$

be a discrete function space defined by linear Lagrange finite elements on a regular triangulation $\mathcal{T}$ of Ω and let $\{\phi_i\}_{i=1}^{n}$ be a basis of $\mathcal{V}_h$. Then inserting an approximation

$$z \approx \sum_{i=1}^{n} \alpha_i \phi_i \in \mathcal{V}_h,$$

into the weak formulation (2.5.6) with the test functions $\psi = \phi_j$, $j = 1, \dots, n$, results in a linear system of equations

$$\mathcal{A} z_h = \ell,$$

where $z_h = [\alpha_1, \dots, \alpha_n]^T$ and the entries of the matrix $\mathcal{A}$ and the vector ℓ are given by

$$\mathcal{A}_{i,j} = b(\phi_i, \phi_j) = \int_\Omega \nabla\phi_i \cdot \nabla\phi_j \, d\xi, \qquad i, j = 1, \dots, n, \tag{2.5.7}$$

$$\ell_j = l(\phi) = \int_\Omega f\phi_j d\xi, \qquad j = 1, \dots, n. \tag{2.5.8}$$

Remark 2.36. Since the bilinear form b in (2.5.5) is symmetric, i.e., $b(v, w) = b(w, v)$ for all $v, w \in \mathcal{V}$, and coercive, the matrix $\mathcal{A}$ is symmetric and positive definite. The latter immediately follows from

$$z^T \mathcal{A} z = \sum_{i=1}^{n} z_i \sum_{j=1}^{n} z_j b(\phi_i, \phi_j) = b\left(\sum_{i=1}^{n} z_i\phi_i, \sum_{j=1}^{n} z_j\phi_j\right) \geqslant m\|\sum_{i=1}^{n} z_i\phi_i\|^2 > 0$$

which holds for all vectors $z \neq 0$.

2.5.3 Nonlinear problems

The FEM can also be extended to nonlinear problems. As an example, we consider a nonlinear Poisson's equation

$$\begin{aligned} \nabla \cdot (a(\nabla z)\nabla z) &= f && \text{in } \Omega, \\ z &= 0 && \text{on } \Gamma_1, \\ \frac{\partial z}{\partial n_0} &= 0 && \text{on } \Gamma_2, \end{aligned}$$

where a is a positive, differentiable function and $a(v)v$ is Lipschitz continuous and strongly monotone, i.e., there exist $L, m > 0$ such that

$$\begin{aligned} \|a(v)v - a(w)w\|_{\mathcal{L}^2(\Omega)} &\leqslant L\|v - w\|_{\mathcal{L}^2(\Omega)}, \\ \langle a(v)v - a(w)w, v - w\rangle &\geqslant m\|v - w\|^2_{\mathcal{L}^2(\Omega)} \end{aligned}$$

for all $v, w \in \mathcal{L}^2(\Omega)$. The variational formulation is determined in the same way as in the linear case and is given by

$$\int_\Omega a(\nabla z)\nabla z \cdot \nabla\psi \, d\xi = \int_\Omega f\psi \, d\xi, \qquad \psi \in \mathcal{D}(\Omega). \tag{2.5.10}$$

We are searching for a weak solution $z \in \mathcal{V}$ satisfying this equation for all $\psi \in \mathcal{D}(\Omega)$, where $\mathcal{D}(\Omega) = \mathcal{V}$ is as in (2.5.2). Introducing

$$b(z, \psi) = \int_\Omega a(\nabla z)\nabla z \cdot \nabla\psi \, d\xi, \tag{2.5.11}$$

$$l(\psi) = \int_\Omega f\psi \, d\xi, \tag{2.5.12}$$

equation (2.5.10) can be written as

$$b(z, \psi) = l(\psi) \qquad \text{for all } \psi \in \mathcal{D}(\Omega). \tag{2.5.13}$$

Since b is not bilinear any more, Theorem 2.32 can not be applied here. This result can, fortunately, be extended to nonlinear problems.

Definition 2.37. Let $\mathcal{H}$ be a Hilbert space. A mapping

$$b : \mathcal{H} \times \mathcal{H} \to \mathbb{R}$$

is called

1. *linear in the second argument* if

$$b(v, \alpha w_1 + \beta w_2) = \alpha b(v, w_1) + \beta b(v, w_2)$$

for all $\alpha, \beta \in \mathbb{R}$ and all $v, w_1, w_2 \in \mathcal{H}$,

2. *Lipschitz continuous in the first argument* if there exists $L > 0$ such that

$$|b(u,w) - b(v,w)| \leqslant L\|u - v\|_{\mathcal{H}}\|w\|_{\mathcal{H}}$$

for all $u, w, v \in \mathcal{H}$,

3. *strongly monotone in the first argument* if there exists $m > 0$ such that

$$b(u, u - v) - b(v, u - v) \geqslant m\|u - v\|^2_{\mathcal{H}}$$

for all $u, v, w \in \mathcal{H}$.

Theorem 2.38 (Zarantonello Theorem). *Let $\mathcal{H}$ be a Hilbert space and $\mathcal{H}'$ be its dual space. Consider a linear continuous functional $l : \mathcal{H} \to \mathbb{R}$ and a mapping $b : \mathcal{H} \times \mathcal{H} \to \mathbb{R}$ which is Lipschitz continuous and strongly monotone in the first argument and linear in the second argument. Then there exists a unique weak solution $v \in \mathcal{H}$ of*

$$b(v, \psi) = l(\psi) \qquad \text{for all } \psi \in \mathcal{D}(\Omega).$$

Proof. See [Zei90b, Theorem 25B]. □

In order to be able to apply Theorem 2.38 to the nonlinear Poisson problem (2.5.13), we have to verify that b and l in (2.5.11) and (2.5.12) satisfy the assumptions of this theorem. Definitions (2.5.12) and (2.5.11) directly show that l is linear and b is linear in the second argument. Furthermore, using the Hölder inequality and the Lipschitz continuity of $a(v)v$, we have

$$\begin{aligned}
|b(u,w) - b(v,w)| &= \left| \int_\Omega a(\nabla u)\nabla u \nabla w - a(\nabla v)\nabla v \nabla w \, d\xi \right| \\
&\leqslant \|a(\nabla u)\nabla u - a(\nabla v)\nabla v\|_{\mathcal{L}^2(\Omega)} \|\nabla w\|_{\mathcal{L}^2(\Omega)} \\
&\leqslant L\|\nabla u - \nabla v\|_{\mathcal{L}^2(\Omega)} \|\nabla w\|_{\mathcal{L}^2(\Omega)} \\
&\leqslant L\|u - v\|_{\mathcal{V}} \|w\|_{\mathcal{V}}
\end{aligned}$$

for all $u, v, w \in \mathcal{V}$. This means that b is Lipschitz continuous in the first argument. The strong monotonicity of b in the first argument follows from the strong monotonicity of $a(u)u$ and Theorem 2.33. Thus, Theorem 2.38 implies the existence and uniqueness of the solution of problem (2.5.13).

Applying the FEM discretization as in Section 2.5.2 to (2.5.13), we obtain a nonlinear system of equations

$$\mathscr{F}(z_h) = 0, \tag{2.5.14}$$

with a nonlinear function $\mathscr{F}(z_h) = \mathscr{A}(z_h)z_h - \ell$, where the entries of the matrix $\mathscr{A}(z_h)$ and the vector ℓ are given by

$$\big(\mathscr{A}(z_h)\big)_{i,j} = \int_\Omega a\Big(\sum_{k=1}^{n} \alpha_k \phi_k\Big) \nabla \phi_i \cdot \nabla \phi_j \, d\xi, \qquad i, j = 1, \ldots, n, \tag{2.5.15a}$$

$$\ell_j = l(\phi_j) = \int_\Omega f \phi_j \, d\xi, \qquad j = 1, \ldots, n. \tag{2.5.15b}$$

Unlike the linear case, we cannot compute the integrals in (2.5.15a) exactly, because of the presence of the nonlinear function a. The entries of $\mathscr{A}$ are determined using numerical integration schemes. The nonlinear equation (2.5.14) is solved by employing the Newton's method, see, for example, [Atk89], which requires the Jacobi matrix of the nonlinear function $\mathscr{F}$. It can be determined using numerical or automatic differentiation [Ral81, WR00].

2.5.4 Time-dependent problems

We now briefly discuss the discretization of time-dependent PDEs. For this purpose, we use the method of lines based on a spatial discretization using FEM followed by time integration of n generated ODEs. For more details on this method and also other techniques for solving time-dependent PDEs, we refer to [Zul11] and [QV94, Chapter 11].

As an example, we consider here a linear heat equation

$$\frac{\partial z}{\partial t} - a\,\triangle z = f \qquad \text{in } [0,T]\times\Omega, \tag{2.5.16a}$$

$$z = 0 \qquad \text{on } [0,T]\times\Gamma_1, \tag{2.5.16b}$$

$$\frac{\partial z}{\partial n_0} = 0 \qquad \text{on } [0,T]\times\Gamma_2, \tag{2.5.16c}$$

$$z(\cdot,0) = z_0 \qquad \text{on } \Omega. \tag{2.5.16d}$$

A weak formulation can be obtained similarly to Poisson's equation in Section 2.5.1. Let the trial space $\mathcal{V}$ be defined as in (2.5.2) and let the test space be given by $\mathcal{D}(\Omega) = \mathcal{V}$. Additionally, we define the spaces

$$\mathcal{L}^2(0,T;\mathcal{V}) = \left\{ v : (0,T) \to \mathcal{V} \;:\; \|v\|^2_{\mathcal{L}^2(0,T;\mathcal{V})} := \int_0^T \|v(t)\|^2_{\mathcal{V}}\, dt < \infty \right\}$$

and

$$H^1(0,T;\mathcal{V}) = \left\{ v \in \mathcal{L}^2(0,T;\mathcal{V}) \;:\; \frac{\partial v}{\partial t} \in \mathcal{L}^2(0,T;\mathcal{V}) \right\}.$$

Then the weak formulation for (2.5.16) has the form

$$\int_\Omega \frac{\partial z}{\partial t}\psi\, d\xi + \int_\Omega a\nabla z\cdot\nabla\psi\, d\xi = \int_\Omega f\,\psi\, d\xi \tag{2.5.17}$$

with $z \in H^1(0,T;\mathcal{V})$ and $\psi \in \mathcal{D}(\Omega)$. Approximating the infinite-dimensional test and trial spaces $\mathcal{V}$ and $\mathcal{D}(\Omega) = \mathcal{V}$ with the finite-dimensional space with the basis $\{\phi_j\}_{j=1}^n$, we insert an approximation

$$z(t,\xi) \approx \sum_{i=1}^{n} \alpha_j(t)\phi_j(\xi)$$

into equation (2.5.17) and test it with the basis functions ϕ_j. As a result, we obtain an ODE system

$$\mathscr{E}\dot{z}_h - \mathscr{A}z_h = \ell, \quad z_h(0) = z_{h0}, \tag{2.5.18}$$

where the entries of the mass matrix $\mathscr{E}$ have the form

$$\mathscr{E}_{i,j} = \int_{\Omega} \phi_i \phi_j d\xi$$

and $\mathscr{A}$ and ℓ are as in (2.5.7) and (2.5.8), respectively. Due to the Piccard-Lindelöff theorem [HNW93, Section I.8], this equation has a unique solution z_h. For numerical solution of (2.5.18), we can use the explicit or implicit one-step or multi-step methods [HNW93, HW96]. As an example, we consider a θ-method. For the linear ODE (2.5.18), this method is given by

$$\mathscr{E} z_{h,k+1} = \mathscr{E} z_{h,k} + \tau(\mathscr{A}(\theta z_{h,k+1} + (1-\theta) z_{h,k}) + \theta\ell((k+1)\tau) + (1-\theta)\ell(k\tau)),$$

where τ is a fixed step size in time, $z_{h,k}$ approximates $z_h(k\tau)$ and $\theta \in [0,1]$. Three methods can be obtained using certain parameters θ: the explicit Euler method for $\theta = 0$, the implicit Euler method for $\theta = 1$ and the midpoint rule for $\theta = \frac{1}{2}$.

3 Numerical solution of coupled systems

In this chapter, we consider the numerical solution of coupled systems of ODEs and DAEs. Such systems arise in many different application areas including micro-electro-mechanical systems and integrated circuit simulation, where different physical effects have to be modeled. Applying spatial discretization of PD(A)Es on complex geometries may also lead to coupled systems, where subsystems are obtained by discretization on subdomains and coupling via boundary conditions. The dynamical behaviour of such systems is characterized by different properties of the interacting subsystems. While powerful methods are available for the different structured subsystems, it is often inefficient to apply these methods to the overall system. A natural approach is to use most suitable simulation algorithms for each subsystem exploiting its structure and properties and couple them in an appropriate way. This approach is known as co-simulation and has been considered in [LBH84, WP99]. For time integration of coupled systems with subsystems having slow and fast dynamic behaviour, multirate methods have been developed in [BG02, CC94, GW84]. Here, we describe the dynamic iteration method, also called *waveform relaxation*, first used for coupled ODEs in [MN92] and then extended to coupled DAEs in [AG01, Ebe08, JK96, PT18].

3.1 Dynamic iteration method

Consider a coupled system of nonlinear DAEs

$$\begin{aligned} f_1(\dot{x}_1, x_1, x_2, \ldots, x_N) &= 0, \qquad x_1(t_0) = x_{10}, \\ &\vdots \\ f_N(\dot{x}_N, x_1, x_2, \ldots, x_N) &= 0, \qquad x_N(t_0) = x_{N0}, \end{aligned} \tag{3.1.1}$$

where $f_j : \mathbb{R}^{n_j} \times \mathbb{R}^{n_1} \times \ldots \times \mathbb{R}^{n_N} \to \mathbb{R}^{n_j}$ are sufficiently smooth functions, $x_{j0} \in \mathbb{R}^{n_j}$ are initial vectors and $x_j : [t_0, T] \to \mathbb{R}^{n_j}$ are unknown functions. We assume that the initial conditions are consistent and system (3.1.1) is solvable. Every subsystem $f_j(\dot{x}_j, x_1, \ldots, x_N) = 0$ can be viewed as a control DAE system with the state vector x_j, the inputs $x_1, \ldots, x_{j-1}, x_{j+1}, \ldots, x_N$ and the output $y_j = x_j$. This allows us to solve the subsystems separately once good approximations to inputs are available. Such approximations can be obtained iteratively using the dynamic iteration method. For this purpose, we split the time interval $[t_0, T]$ into n_T macro time windows $[T_m, T_{m+1}]$ with a time grid $t_0 = T_0 < T_1 < \ldots < T_{n_T} = T$. Then the subsystems are solved iteratively on every time window using the data from the previous iteration steps and, if necessary, from the previous time windows. A dynamic iteration step on the time window $[T_m, T_{m+1}]$ is given by

$$\Sigma_{m,k}: \begin{cases} f_1(\dot{x}_1^{[k]}, X_{1,1}^{[k]}, X_{1,2}^{[k]}, \dots, X_{1,N}^{[k]}) = 0, & x_1^{[k]}(T_m) = \hat{x}_1^{[K_{m-1}]}(T_m), \\ \vdots & \\ f_N(\dot{x}_N^{[k]}, X_{N,1}^{[k]}, X_{N,2}^{[k]}, \dots, X_{N,N}^{[k]}) = 0, & x_N^{[k]}(T_m) = \hat{x}_N^{[K_{m-1}]}(T_m), \end{cases} \tag{3.1.2}$$

for $k = 1, \dots, K_m$ with

$$X_{i,j}^{[k]} = \sum_{l=0}^{\ell} W_{i,j}^{[l]} x_j^{[k-l]}, \qquad i, j = 1, \dots, N,$$

where $W_{i,j}^{[l]} \in \mathcal{C}^0(\mathbb{I}, \mathbb{R}^{n_j \times n_j})$ are appropriately chosen weights, $x_j^{[k]}$ is the k-th iterate of the j-th subsystem and $\hat{x}_i^{[K_{m-1}]}(T_m)$ is the last iterate on the time window $[T_{m-1}, T_m]$ evaluated at T_m or, if $m = 0$, the initial value x_{j0}. In most cases, the weights $W_{i,j}^{[l]}$ are time-independent. The value ℓ is called the depth of the method. For reasons of simplicity, we do not mark the macro time step on the variable $x_j^{[k]}$ and only consider it on the appropriate time interval or extrapolate it to the next macro time step.

Figure 3.1 shows the data flow for the dynamic iteration method. The macro time windows $[T_m, T_{m+1}]$ are given on the horizontal axis, whereas the vertical axis indicates the iteration steps. On every macro time interval $[T_m, T_{m+1}]$, we use the data $x_1^{[K_{m-1}]}, \dots, x_N^{[K_{m-1}]}$ from the last iteration K_{m-1} on the previous macro time step to define the initial values and the first approximation to the inputs, this is marked as "initialize" in Figure 3.1. For DAEs, we have also to ensure the consistency of the initial values. Then we iterate by integrating the subsystems until the approximated solutions satisfy a given tolerance and go on to the next macro window. At the end of the dynamic iteration, we have at any time T_m data for all subsystems. Within the macro time steps, we solve the subsystems on separate fine time grids using appropriate integration methods.

Depending on the choice of $W_{i,j}^{[k]}$, the dynamic iteration has different properties. Next, we consider the Jacobi- and Gauß-Seidel-type iterations.

Definition 3.1. The dynamic iteration method is of *Jacobi-type* if the depth $\ell = 1$ and

$$W_{i,j}^{[0]} = \begin{cases} I, & i = j \\ 0, & i \neq j \end{cases}, \qquad W_{i,j}^{[1]} = \begin{cases} 0, & i = j \\ I, & i \neq j \end{cases}.$$

The dynamic iteration method is of *Gauß-Seidel-type* if the depth $\ell = 1$ and

$$W_{i,j}^{[0]} = \begin{cases} 0, & j > i \\ I, & j \leqslant i \end{cases}, \qquad W_{i,j}^{[1]} = \begin{cases} I, & j > i \\ 0, & j \leqslant i \end{cases}.$$

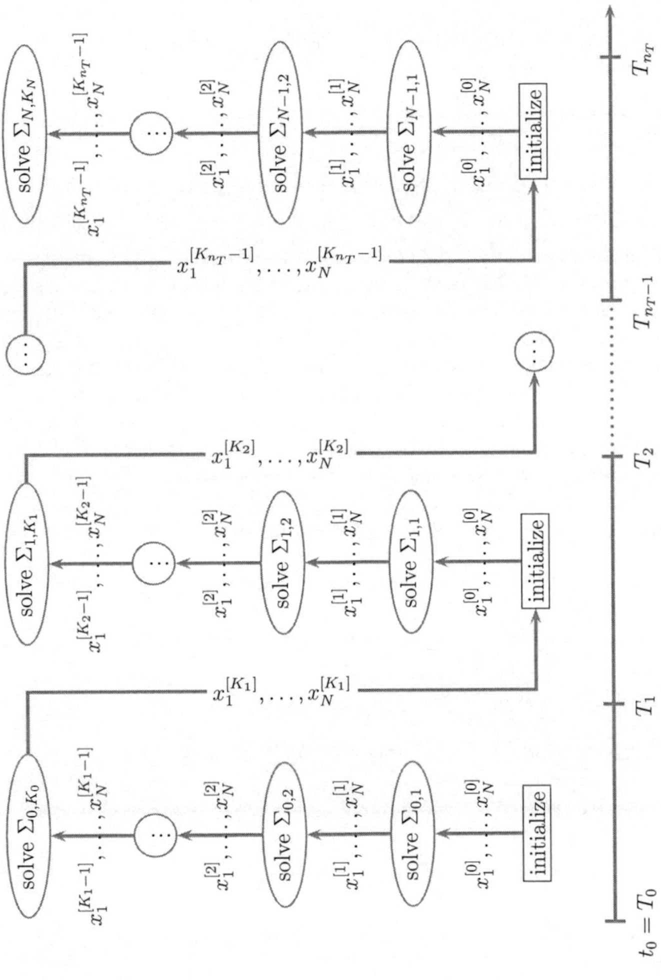

Figure 3.1: The data flow diagram for the dynamic iteration method

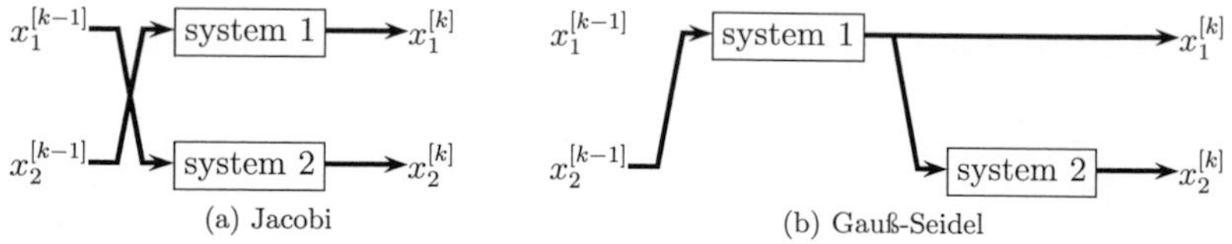

Figure 3.2: The data flow diagram for the Jacobi-type and Gauß-Seidel-type dynamic iteration step

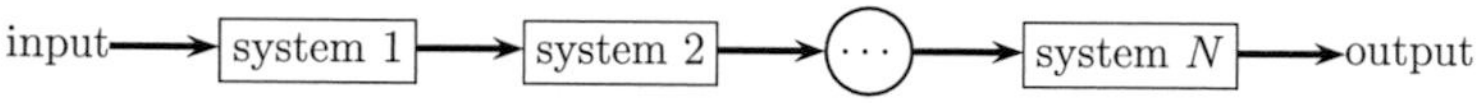

Figure 3.3: Sequential coupled system

Figure 3.2(a) shows the data flow in the Jacobi-type dynamic iteration step for two subsystems. This approach has the advantage that it can easily be parallelized, since all subsystems can be solved simultaneously. Figure 3.2(b) presents the Gauß-Seidel-type dynamic iteration step for two subsystems. This method is sequential that can be a disadvantage for some special systems. Which of these methods should be used, depends on the coupling structure of the coupled system. For example, solving a coupled system indicated in Figure 3.3 with the Jacobi-type method requires N iterations of every subsystem on the time window $[T_m, T_{m+1}]$, whereas in the Gauß-Seidel-type method, every subsystem is solved only once.

In the following, we present some theoretical results on the convergence of the dynamic iteration method. First, we consider only ODEs.

Theorem 3.2. *Suppose that the coupled ODE system*

$$\begin{aligned} \dot{x}_1 &= f_1(x_1, \ldots, x_N), \quad x_1(t_0) = x_{10}, \\ &\vdots \\ \dot{x}_N &= f_N(x_1, \ldots, x_N),\ x_N(t_0) = x_{N0}, \end{aligned} \tag{3.1.3}$$

is uniquely solvable on $[t_0, T]$. *The dynamic iteration method*

$$\begin{aligned} \dot{x}_1^{[k]} &= f_1(X_{1,1}^{[k]}, X_{1,2}^{[k]}, \ldots, X_{1,N}^{[k]}), \quad x_1^{[k]}(t_0) = x_{10}, \\ &\vdots \\ \dot{x}_N^{[k]} &= f_N(X_{N,1}^{[k]}, X_{N,2}^{[k]}, \ldots, X_{N,N}^{[k]}),\ x_N^{[k]}(t_0) = x_{N0}, \end{aligned}$$

converges for $k \to \infty$ *to the solution of* (3.1.3) *if the starting iterates* $x_1^{[0]}, \ldots, x_N^{[0]}$ *are continuous and*

$$\sum_{l=0}^{\ell} W_{i,j}^{[l]} = I, \quad i, j = 1, \ldots, N. \tag{3.1.4}$$

Proof. See [Ebe08, Theorem 3.1.3]. □

Note that both the Jacobi-type and the Gauß-Seidel-type dynamic iterations satisfy the condition (3.1.4). Unfortunately, Theorem 3.2 does not hold in general for DAEs. For simplicity, we restrict ourselves to the coupled DAE system (3.1.1) of differentiation index at most 1. We also assume that all subsystems fulfill this index condition. Moreover, we consider the dynamic iteration method (3.1.2) of depth $l = 1$. In this case, (3.1.2) can be written as $F(\dot{x}^{[k]}, x^{[k]}, x^{[k-1]}) = 0$ with $x^{[k]} = \left[\left(x_1^{[k]}\right)^T, \ldots, \left(x_N^{[k]}\right)^T\right]^T \in \mathbb{R}^n$, $n = n_1 + \ldots + n_N$. Assume that this DAE system has differentiation index 1 with respect to $x^{[k]}$. Then differentiating it, we can extract the underlying ODE

$$\dot{x}^{[k]} = \varphi(x^{[k]}, x^{[k-1]}, \dot{x}^{[k-1]}), \tag{3.1.5}$$

The following theorem is an extension of Theorem 3.2 to the coupled DAE system (3.1.1) satisfying the above conditions.

Theorem 3.3. *Consider a coupled DAE system* (3.1.1) *such that the differentiation index of each subsystem, the coupled system itself and the corresponding dynamic iteration system* (3.1.2) *is at most 1. Let* (3.1.5) *be the underlying ODE of system* (3.1.2) *with* $\varphi \in \mathcal{C}^2(\mathbb{R}^n \times \mathbb{R}^n \times \mathbb{R}^n, \mathbb{R}^n)$ *and let* $x^{[k]} \in \mathcal{C}^1([t_0, T], \mathbb{R}^n)$ *be the solution of* (3.1.2) *with an initial condition* $x^{[k]}(t_0) = x_0$ *and a starting iterate* $x^{[0]} \in \mathcal{C}^1([t_0, T], \mathbb{R}^n)$. *If a constant* $L < 1$ *exists such that*

$$\left\| \frac{\partial \varphi}{\partial \dot{x}^{[k-1]}}(\zeta^{[k]}, \zeta^{[k-1]}, \xi^{[k-1]}) \right\|_\infty \leqslant L$$

for arbitrary $\zeta^{[k]}$, $\zeta^{[k-1]}$, $\xi^{[k-1]} \in \mathbb{R}^n$, *then* $x^{[k]}$ *converges for* $k \to \infty$ *to a solution of the coupled system.*

Proof. See [Ebe08, Theorem 3.2.6]. □

3.2 Dynamic iteration using reduced-order models

A combination of dynamic iteration and model order reduction called *dynamic iteration using reduced-order models* (DIRM) was introduced in [RP02]. The main idea of this approach is to use MOR to generate the input variables for the subsystems that are otherwise determined by interpolation or extrapolation. This has the advantage that the inputs become dependent on the current solution. In nonlinear case, the reduced-order models are computed by POD or POD-DEIM based on snapshots as discussed in Sections 2.4.2 and 2.4.3. Since the generation of snapshots is usually expensive, one uses the previous iterates $x_j^{[l]}$, $l \leqslant k$, to generate the POD and POD-DEIM basis matrices at the k-th iteration step.

Consider again the coupled ODE system (3.1.3). Let

$$T_m \leqslant t_{m,1} < \ldots < t_{m,s_m} \leqslant T_{m+1}$$

be time instances on the time window $[T_m, T_{m+1}]$ and let

$$\mathcal{X}^{[k]} = \{x_j^{[l]}(t_{m,1}), \ldots, x_j^{[l]}(t_{m,s_m}),\ j = 1, \ldots, N,\ l = 0, \ldots, k\}$$

and

$$\mathcal{X}_f^{[k]} = \Big\{ f_j\big(x_1^{[l]}(t_{m,1}), \dots, x_N^{[l]}(t_{m,1})\big), \dots, \\ f_j\big(x_1^{[l]}(t_{m,m_s}), \dots, x_N^{[l]}(t_{m,m_s})\big), \quad j = 1, \dots, N,\, l = 0, \dots, k\Big\}$$

be the set of snapshots for $x = [x_1^T, \dots, x_N^T]^T$ and $f = [f_1^T, \dots, f_N^T]^T$. We are using the selectors $\Xi_{i,j}^{[k]}$ and $\Theta_{i,j}^{[k]}$, where $i = 1, \dots, N$ is the index of the subsystem, $j = 1, \dots, N$, $j \neq i$, is the index of the variable to be reduced, and $k = 1, \dots, K_m$ is the iteration index. We determine from $\Xi_{i,j}^{[k]}\mathcal{X}^{[k]}$ the POD basis matrices $V_{i,j}^{[k]}$ and from $\Theta_{i,j}^{[k]}\mathcal{X}_f^{[k]}$ the DEIM basis and selector matrices $U_{i,j}^{[k]}$ and $\mathcal{S}_{i,j}^{[k]}$, respectively. Then the k-th iteration step of the DIRM on the time window $[T_m, T_{m+1}]$ is given by N coupled systems

$$\begin{aligned}
\dot{\hat{x}}_{i,1}^{[k]} &= \hat{f}_{i,1}^{[k]}(\hat{X}_{i,1}^{[k]}, \dots, \hat{X}_{i,i-1}^{[k]}, x_i^{[k]}, \hat{X}_{i,i+1}^{[k]}, \dots, \hat{X}_{i,N}^{[k]}),\\
&\vdots\\
\dot{\hat{x}}_{i,i-1}^{[k]} &= \hat{f}_{i,i-1}^{[k]}(\hat{X}_{i,1}^{[k]}, \dots, \hat{X}_{i,i-1}^{[k]}, x_i^{[k]}, \hat{X}_{i,i+1}^{[k]}, \dots, \hat{X}_{i,N}^{[k]}),\\
\dot{x}_i^{[k]} &= f_i(\hat{X}_{i,1}^{[k]}, \dots, \hat{X}_{i,i-1}^{[k]}, x_i^{[k]}, \hat{X}_{i,i+1}^{[k]}, \dots, \hat{X}_{i,N}^{[k]}),\\
\dot{\hat{x}}_{i,i+1}^{[k]} &= \hat{f}_{i,i+1}^{[k]}(\hat{X}_{i,1}^{[k]}, \dots, \hat{X}_{i,i-1}^{[k]}, x_i^{[k]}, \hat{X}_{i,i+1}^{[k]}, \dots, \hat{X}_{i,N}^{[k]}),\\
&\vdots\\
\dot{\hat{x}}_{i,N}^{[k]} &= \hat{f}_{i,N}^{[k]}(\hat{X}_{i,1}^{[k]}, \dots, \hat{X}_{i,i-1}^{[k]}, x_i^{[k]}, \hat{X}_{i,i+1}^{[k]}, \dots, \hat{X}_{i,N}^{[k]}),
\end{aligned} \tag{3.2.1}$$

with

$$\hat{f}_{i,j}^{[k]}(\hat{X}_{i,1}^{[k]}, \dots, \hat{X}_{i,i-1}^{[k]}, x_i^{[k]}, \hat{X}_{i,i+1}^{[k]}, \dots, \hat{X}_{i,N}^{[k]}) \\ = Z_{i,j}^{[k]} f_j(\hat{X}_{i,1}^{[k]}, \dots, \hat{X}_{i,i-1}^{[k]}, x_i^{[k]}, \hat{X}_{i,i+1}^{[k]}, \dots, \hat{X}_{i,N}^{[k]})$$

and

$$\begin{aligned}
\hat{X}_{i,j}^{[k]} &= V_{i,j}^{[k]}\hat{x}_{i,j}^{[k]}, & j = 1, \dots, N,\, j \neq i,\\
Z_{i,j}^{[k]} &= \left(V_{i,j}^{[k]}\right)^T U_{i,j}^{[k]}\left(\left(\mathcal{S}_{i,j}^{[k]}\right)^T U_{i,j}^{[k]}\right)^{-1}\left(\mathcal{S}_{i,j}^{[k]}\right)^T, & j = 1, \dots, N,\, j \neq i
\end{aligned}$$

for $i = 1, \dots, N$. Here $\hat{x}_{j,i}^{[k]} \in \mathbb{R}^{\eta_{i,j}}$ are the reduced variables and $x_i^{[k]}$ the DIRM iterate. One can see that the DIRM has the same work flow as the dynamic iteration presented in Figure 3.1. The only difference is that instead of solving N subsystems of dimension n_i each, in the DIRM, we solve N systems of dimension $n_i + \sum_{j=1, j\neq i}^{N} \eta_{i,j}$ each, where $\eta_{i,j}$ is the dimension of the POD basis $V_{i,j}^{[k]}$ for every subsystem. The additional computation effort will, hopefully, be compensated by speeding up the convergence, since using the reduced models provides more accurate approximations to the input variables for the subsystems. In Figure 3.4, we collect the variables and functions for the dynamic iteration and DIRM and explain different indices.

The different selectors $\Xi_{i,j}^{[k]}$ and $\Theta_{i,j}^{[k]}$ lead to different reduced-order models. Here, we present two possible choices:

- Jacobi-type selector

$$\Xi_{i,j}^{[k]}\mathcal{X}^{[k]} = \{x_j^{[k-1]}(t_{m,1}), \dots, x_j^{[k-1]}(t_{m,s_m})\}$$

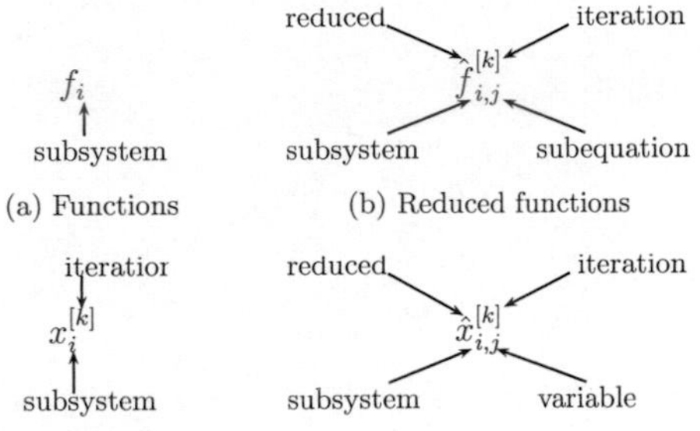

(a) Functions (b) Reduced functions (c) Iterates (d) Reduced iterates

Figure 3.4: Functions and iterate variables for the dynamic iteration and DIRM

- Gauß-Seidel-type selector

$$\Xi_{i,j}^{[k]} \mathcal{X}^{[k]} = \begin{cases} \{x_j^{[k-1]}(t_{m,1}), \ldots, x_j^{[k-1]}(t_{m,s_m})\}, & \text{for } i < j, \\ \{x_j^{[k]}(t_{m,1}), \ldots, x_j^{[k]}(t_{m,s_m})\}, & \text{for } i > j, \end{cases}$$

for every DIRM subsystem i and every iteration k. Note that for the Jacobi-type selector, the variables from the previous iteration step are taken as snapshots for all subsystems. In this case, $\Xi_{i,j}^{[k]}$ does not actually depend on i. In contrast to this, for the Gauß-Seidel-type selector $\Xi_{i,j}^{[k]}$ the POD basis $V_{i,1}^{[k]}, \ldots, V_{i,i-1}^{[k]}$ for the i-th DIRM system (3.2.1) are determined from the actual snapshots of the subsystems $1, \ldots, i-1$, whereas the POD basis $V_{i,i+1}^{[k]}, \ldots, V_{i,N}^{[k]}$ are obtained from the snapshots from the previous iteration. The selectors $\Theta_{i,j}^{[k]}$ for the functions f_j can be chosen in a similar way. Note that starting from the second iteration, we can extend the snapshot set $\mathcal{X}_f^{[k]}$ by additional information

$$f_j\left(\hat{X}_{1,j}^{[l]}(t_{m,1}), \ldots, \hat{X}_{j-1,j}^{[l]}(t_{m,1}), x_j^{[l]}(t_{m,1}), \hat{X}_{j,j+1}^{[l]}(t_{m,1}), \ldots, \hat{X}_{N,j}^{[l]}(t_{m,1})\right),$$
$$\vdots$$
$$f_j\left(\hat{X}_{1,j}^{[l]}(t_{m,s_m}), \ldots, \hat{X}_{j-1,j}^{[l]}(t_{m,s_m}), x_j^{[l]}(t_{m,s_m}), \hat{X}_{j,j+1}^{[l]}(t_{m,s_m}), \ldots, \hat{X}_{N,j}^{[l]}(t_{m,s_m})\right)$$

available from the previous steps. We have to make an initial step to get first snapshots. We decided here to perform one step of Jacobi-type dynamic iteration. Extrapolation gives only linear dependent vectors and does not extend the snaphot matrix.

In contrast to the dynamic iteration method, the DIRM is not well studied yet. In [RP02], the convergence of the DIRM is investigated for linear systems only. Here, we briefly overview these results. Consider a coupled linear system with two subsystems

$$\begin{aligned} \dot{x}_1 &= A_{11}x_1 + A_{12}x_2, & x_1(t_0) &= x_{10}, \\ \dot{x}_2 &= A_{21}x_1 + A_{22}x_2, & x_2(t_0) &= x_{20}, \end{aligned} \tag{3.2.2}$$

where $A_{ij} \in \mathbb{R}^{n_i \times n_j}$ for $i, j = 1, 2$. Then the Jacobi-type DIRM subsystems take the form

$$\begin{aligned}\dot{x}_1^{[k]} &= A_{11}x_1^{[k]} + A_{12}V_{1,2}^{[k]}\hat{x}_{1,2}^{[k]},\\ \dot{\hat{x}}_{1,2}^{[k]} &= \left(V_{1,2}^{[k]}\right)^T A_{21}x_1^{[k]} + \left(V_{1,2}^{[k]}\right)^T A_{22}V_{1,2}^{[k]}\hat{x}_{1,2}^{[k]}\end{aligned}$$

and

$$\begin{aligned}\dot{\hat{x}}_{2,1}^{[k]} &= \left(V_{2,1}^{[k]}\right)^T A_{11}V_{2,1}^{[k]}\hat{x}_{2,1}^{[k]} + \left(V_{2,1}^{[k]}\right)^T A_{12}x_2^{[k]},\\ \dot{x}_2^{[k]} &= A_{21}V_{2,1}^{[k]}\hat{x}_{2,1}^{[k]} + A_{22}x_2^{[k]}\end{aligned}$$

with the projection matrices $V_{1,2}^{[k]} \in \mathbb{R}^{n_1 \times \eta_2}$ and $V_{2,1}^{[k]} \in \mathbb{R}^{n_2 \times \eta_1}$ such that $\left(V_{2,1}^{[k]}\right)^T V_{2,1}^{[k]} = I_{\eta_2}$ and $\left(V_{1,2}^{[k]}\right)^T V_{1,2}^{[k]} = I_{\eta_1}$. Introducing the orthogonal projectors $P_1^{[k]} = V_{2,1}^{[k]}\left(V_{2,1}^{[k]}\right)^T$ and $P_2^{[k]} = V_{1,2}^{[k]}\left(V_{1,2}^{[k]}\right)^T$, these subsystems can be written as

$$\begin{aligned}\dot{x}_1^{[k]} &= A_{11}x_1^{[k]} + A_{12}z_2^{[k]},\\ \dot{z}_2^{[k]} &= P_2^{[k]}A_{21}x_1^{[k]} + P_2^{[k]}A_{22}z_2^{[k]}\end{aligned} \tag{3.2.3}$$

and

$$\begin{aligned}\dot{z}_1^{[k]} &= P_1^{[k]}A_{11}z_1^{[k]} + P_1^{[k]}A_{12}x_2^{[k]},\\ \dot{x}_2^{[k]} &= A_{21}z_1^{[k]} + A_{22}x_2^{[k]},\end{aligned} \tag{3.2.4}$$

where $z_1^{[k]} = V_{2,1}^{[k]}\hat{x}_{2,1}^{[k]}$ and $z_2^{[k]} = V_{1,2}^{[k]}\hat{x}_{1,2}^{[k]}$ are approximations to x_1 and x_2, respectively. To specify the dynamic iteration operator, we introduce

$$P^{[k]} = \begin{bmatrix} P_1^{[k]} & 0 \\ 0 & P_2^{[k]} \end{bmatrix}, \qquad A_d = \begin{bmatrix} A_{11} & 0 \\ 0 & A_{22} \end{bmatrix}, \qquad A_0 = \begin{bmatrix} 0 & A_{12} \\ A_{21} & 0 \end{bmatrix}$$

and write (3.2.3) and (3.2.4) together as

$$\begin{aligned}\dot{x}^{[k]} &= A_d x^{[k]} + A_0 z^{[k]},\\ \dot{z}^{[k]} &= P^{[k]}A_d x^{[k]} + P^{[k]}A_0 z^{[k]}\end{aligned}$$

with

$$x^{[k]} = \begin{bmatrix} x_1^{[k]} \\ x_2^{[k]} \end{bmatrix}, \qquad z^{[k]} = \begin{bmatrix} z_1^{[k]} \\ z_2^{[k]} \end{bmatrix}.$$

Define the iteration operator $\mathcal{I} : \mathcal{L}^2(0, T; \mathbb{R}^{n_1+n_2}) \to \mathcal{L}^2(0, T; \mathbb{R}^{n_1+n_2})$ which maps the iterate $x^{[k]}$ to $x^{[k+1]}$. Note that the operator $\mathcal{I}$ is nonlinear because POD reduction is a nonlinear mapping. This can be demonstrated by the following example.

Example 3.4. Computing the SVD

$$\begin{bmatrix} 1 & 0 \\ 0 & 0 \end{bmatrix} = \begin{bmatrix} 1 & 0 \\ 0 & 1 \end{bmatrix}\begin{bmatrix} 1 & 0 \\ 0 & 0 \end{bmatrix}\begin{bmatrix} 1 & 0 \\ 0 & 1 \end{bmatrix},$$

we obtain the POD basis $\begin{bmatrix}1 & 0\end{bmatrix}^T$ with reduced dimension 1. Analogously, the SVD

$$\begin{bmatrix}0 & 0\\0 & 1\end{bmatrix}=\begin{bmatrix}0 & 1\\1 & 0\end{bmatrix}\begin{bmatrix}1 & 0\\0 & 0\end{bmatrix}\begin{bmatrix}0 & 1\\1 & 0\end{bmatrix}$$

provides the POD basis $\begin{bmatrix}0 & 1\end{bmatrix}^T$ of dimension 1. However, if we apply POD to

$$a\begin{bmatrix}1 & 0\\0 & 0\end{bmatrix}+b\begin{bmatrix}0 & 0\\0 & 1\end{bmatrix}=\begin{bmatrix}1 & 0\\0 & 1\end{bmatrix}\begin{bmatrix}a & 0\\0 & b\end{bmatrix}\begin{bmatrix}1 & 0\\0 & 1\end{bmatrix}$$

with $a > b > 0$, we get $\begin{bmatrix}1 & 0\end{bmatrix}^T$. A linear operator should result in $\begin{bmatrix}a & b\end{bmatrix}^T$.

The following results on the convergence of the DIRM-method were presented in [RP02]:

- The nonlinear operator $\mathcal{I}$ has a fixed point x^* if $\|A_0\|_2$ is sufficiently small. Unfortunately, no quantitative bound on $\|A_0\|_2$ is given and the proof does not suggest any. This result holds for arbitrary (finite) number of linear subsystems.
- An upper bound on the error of the fixed point trajectory x^* and the through solution $x = \left[x_1^T, x_2^T\right]^T$ of (3.2.2) is presented.
- Assuming that the initial iterate $x^{[0]}$ is sufficiently close to the fixed point x^* of $\mathcal{I}$ and $\|\mathcal{DI}(x^*)\|_2 < 1$, where $\mathcal{DI}(x^*)$ denotes a linearization of the iteration operator $\mathcal{I}$ at x^*, the convergence of the DIRM iteration to x^* is proven.
- Finally, for a coupled linear system, for which the operator $\mathcal{I}$ has a fixed point x^*, the convergence of DIRM is established on a sufficiently small time interval $[t_0, t_0 + \tau]$ under assumption that the initial iterate $x^{[0]}$ is close enough to x^*.

These convergence results cannot be used in practice, since the conditions guaranteeing the convergence are very difficult to verify even for linear systems. An extension to nonlinear coupled systems remains still open.

3.3 Numerical experiments for a nonlinear coupled system

The DIRM method depends on many parameters that strongly influence the convergence of this method. These are, for example, the number of macro steps n_T and the number of iterations K_m on the macro time window $[T_m, T_{m+1}]$. Applying model reduction, we have to choose the reduced dimensions $\eta_{i,j}$ and $\kappa_{i,j}$ for POD and DEIM, respectively, or tolerances as described in (2.4.11). Furthermore, we have to select snapshots required for the construction of the reduced bases, which can be selected by $\Xi_{i,j}^{[k]}$ and $\Theta_{i,j}^{[k]}$. In this section, we present some results of numerical experiments for a simple nonlinear coupled system to demonstrate the difficulties in choosing all these method parameters.

We consider a nonlinear PDE

$$\frac{\partial u}{\partial t} = \nu \frac{\partial^2 u}{\partial \xi^2} + a \frac{\partial u}{\partial x} u$$

with the boundary conditions

$$u(t, 0) = u(t, 6) = 0$$

and the initial conditions

$$u(0, x) = \exp\left(-\left(\frac{x}{3} - 1 \right)^2 \right) \sin\left(\frac{\pi}{2} x \right).$$

for $\xi \in [0, 6]$ and $t \in [0, T]$. For simplicity, we take $\nu = 1$ and $a = 1$. In order to obtain a coupled dynamical system as in (3.2.1), we divide the spatial interval $[0, 6]$ into four equidistant subintervals and discretize (3.3) on each subinterval using a finite difference method [Smi85] with 250 equidistant discretization points. This leads to a nonlinear coupled system with four subsystems of dimension $n_j = 250$, $j = 1, \ldots, 4$, connected sequentially. In the first experiment, we set $T = 1$ and vary only the number K_1 of dynamic iterations on one macro time window $[0, 1]$ and the POD reduced dimension η, which is the same for all iterations and all subsystems. Other parameters collected in Figure 3.5(a) remain fixed.

Note that we do not apply DEIM for the reduction of nonlinearity, since in this small example, the simulation time is so short, that the advantage of DEIM does not come to bear. As a reference solution $x_{\text{ref}} \in \mathbb{R}^{1000}$, we use the solution obtained by discretizing (3.3) on the whole interval $[0, 6]$ using the finite difference method and solving the resulting ODE employing the MATLAB function ode15s with step size control.

In Figure 3.5(b), we present the relative error

$$\Delta_r(T) := \frac{\|x_{\text{ref}}(T) - x(T)\|}{\|x_{\text{ref}}(T)\|}$$

in the DIRM solution after the first two iterations for different POD reduced dimensions. Note that the relative error can only be computed for the time instants T_j for $j = 1, \ldots, n_T$. Since we use $n_T = 1$, we can compare the solutions at time T only. One can see that for increasing dimension η, the relative error decreases. This is a typical behavior of the POD method. It should, however, be noted that the error also depend on the snapshots that are different in each iteration. Moreover, Figure 3.5(b) shows that the second iteration leads to the significantly smaller relative error $\Delta_r(T)$ than the first iteration. For dynamic iteration, this is a basic property, but it is not guaranteed for DIRM.

In Figure 3.5(c), we present the simulation time. Here, DIRM behaves again as expected, the larger the reduced dimension, the longer the simulation time. It is striking to note that the second iteration is significantly more expensive than the first one. We suspect that this is because a POD approximation creates an inappropriate reduced-order subsystem. This leads to long simulation time. Since we want to compute a solution with a small error in a short time, we look at Figure 3.5(d) showing the relative error versus the computation time and search for a point near the origin. Depending on the desired accuracy, a suitable POD dimension η and a number of iterations K_1 can be read out from Figure 3.5(b). From the outset, however, it is nontrivial to select these parameters. The same dimension η can lead to large errors in the first iteration and to long simulation time in the second iteration.

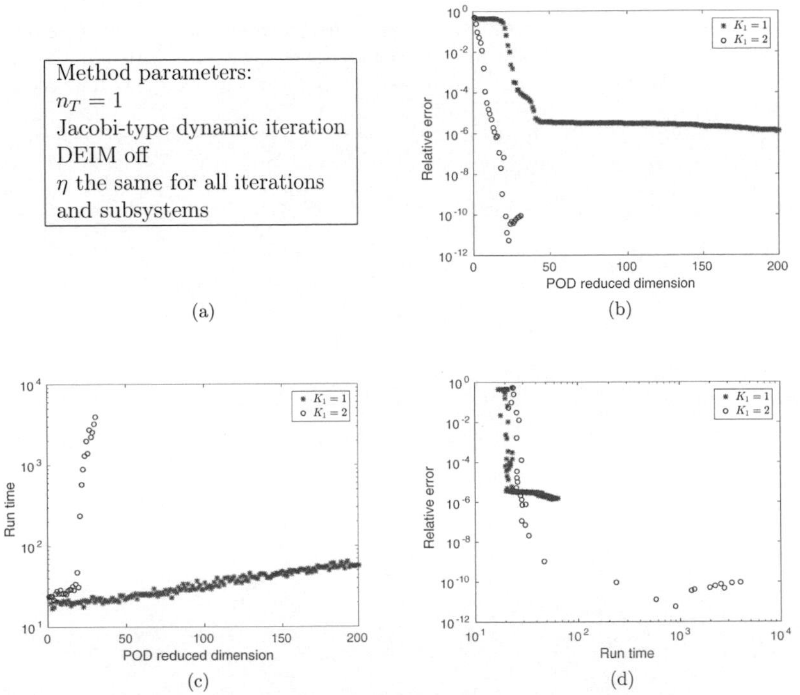

Figure 3.5: Influence of POD dimensions and number of dynamic iterations on the relative error and run time for the Jacobi-type DIRM method

Instead of setting the reduced dimension for POD, one can also select a tolerance *tol* for the POD singular values. We now look at how DIRM reacts to the reduced systems determined by different tolerances. This time, we use the same tolerance $tol = 2^{-i}$ for all subsystems and iteration steps. The resulting POD dimensions for four subsystems are presented in Figure 3.7 for the first and second iterations. It can be seen that in the second iteration, the differences between the subsystems are larger than in the first iteration, which is due to different snapshots for each subsystem. All in all, the reduced dimension increases with decreasing tolerance. In Figures 3.6(b) and 3.6(c), we present the relative errors and simulation time for different tolerances and different number of dynamic iterations. Figure 3.6(d) shows the relative error versus the simulation time. One can see that DIRM with varying tolerances provides about the same results as with varying POD reduced dimensions.

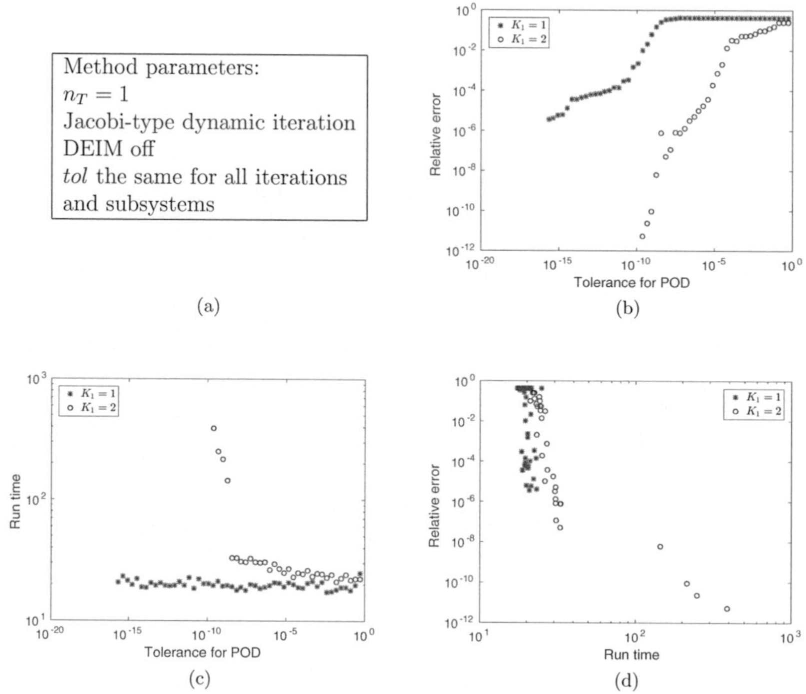

Figure 3.6: Influence of POD tolerance and number of iterations on the relative error and run time for the Jacobi-type DIRM method

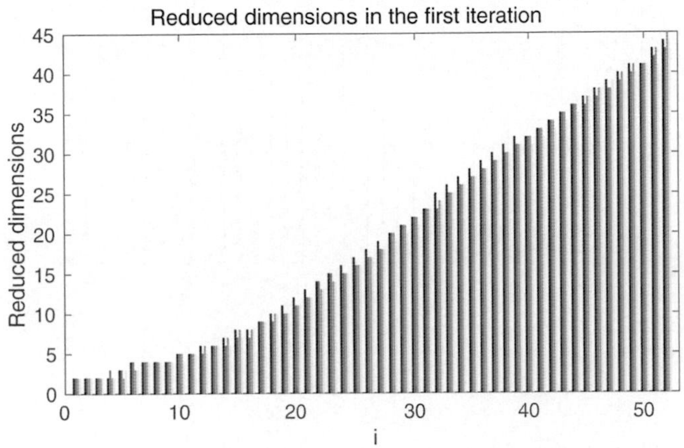

(a)

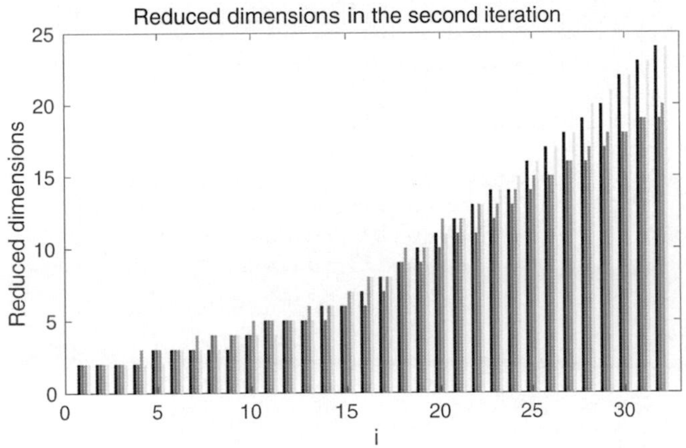

(b)

Figure 3.7: POD reduced dimensions for four subsystems generated with tolerance $tol = 2^{-i}$ in the Jacobi-type DIRM method

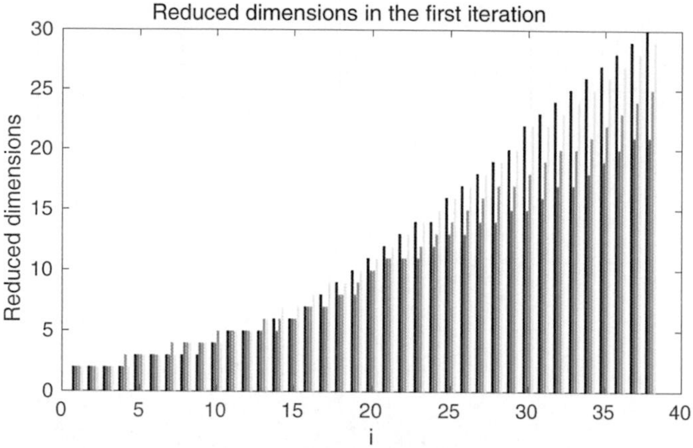

Figure 3.8: POD reduced dimensions for four subsystems generated with tolerance $tol = 2^{-i}$ in the Gauß-Seidel-type DIRM method

In the next set of experiments, we employ the Gauß-Seidel-type dynamic iteration and repeat above computations. While in the Jacobi-type DIRM method, all subsystems are reduced at the beginning of each iteration and then all N coupled systems (3.2.1) are solved in parallel, the Gauß-Seidel-type DIRM iteration is serially in nature. This means that after solving the $(i-1)$-st system (3.2.1), its solution is used to determine the POD basis $V_{i,i-1}^{[k]}$ required in the i-th system. This results in the different choice of dimensions, compare Figures 3.7 and 3.8. If we compare the error plots in Figures 3.6(b) and 3.9(b), we see that the error in the Gauß-Seidel-type DIRM behaves similar to that in the first Jacobi-type iteration. However, comparing the run time in Figures 3.6(c) and 3.9(c), one can see that the simulation time for the Gauß-Seidel-type DIRM is about the same as that for two Jacobi-type iterations. Overall, the Jacobi-type DIRM in this example is more efficient than the Gauß-Seidel-type DIRM.

Method parameters:
$n_T = 1$
Gauß-Seidel-type dynamic iteration
DEIM off
tol is the same for all subsystems
one iteration

(a)

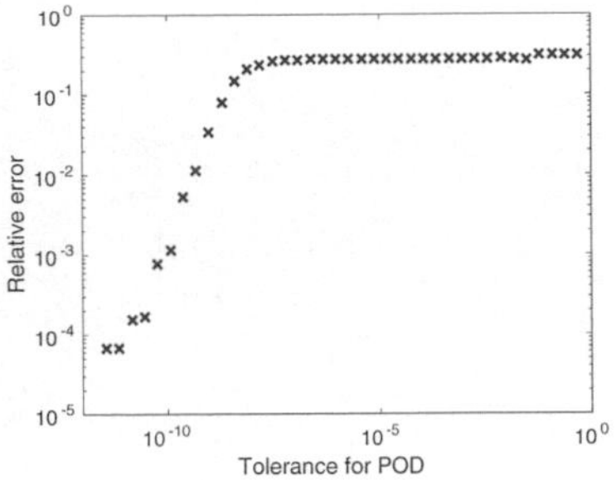

(b)

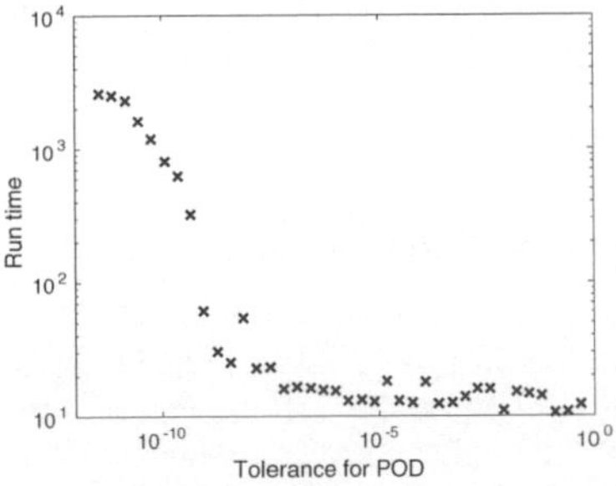

(c)

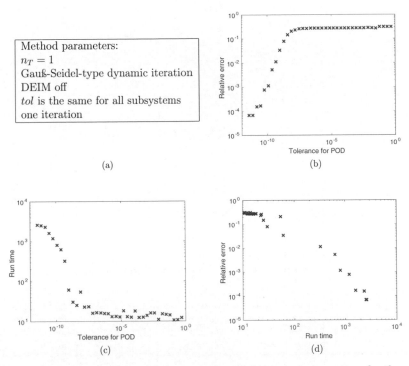

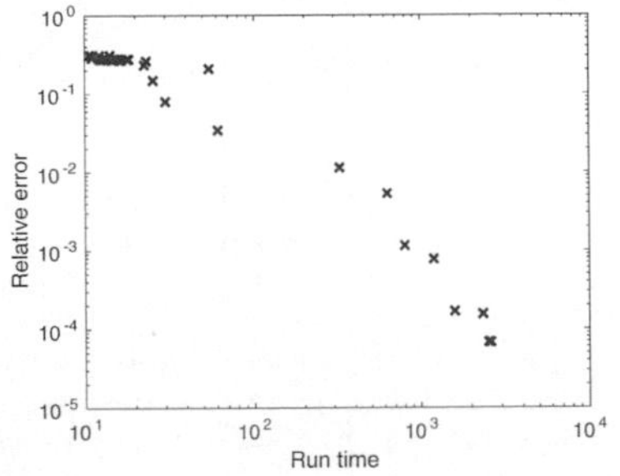

(d)

Figure 3.9: Influence of POD tolerance on the relative error and run time for the Gauß-Seidel-type DIRM method

Method parameters:
Jacobi-type dynamic iteration
DEIM off
tol is the same for all iterations and subsystems
2 iterations

(a)

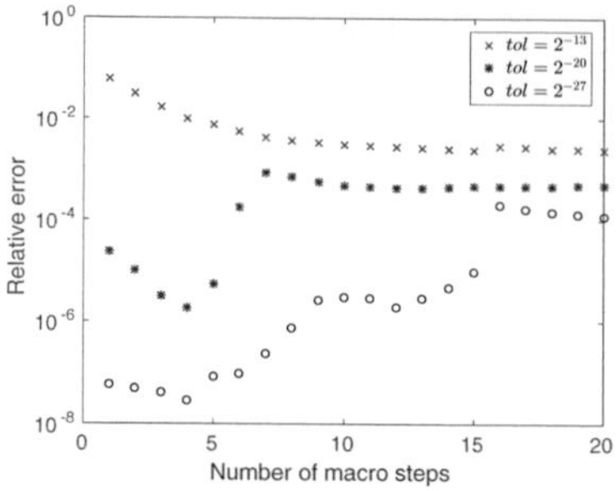

(b)

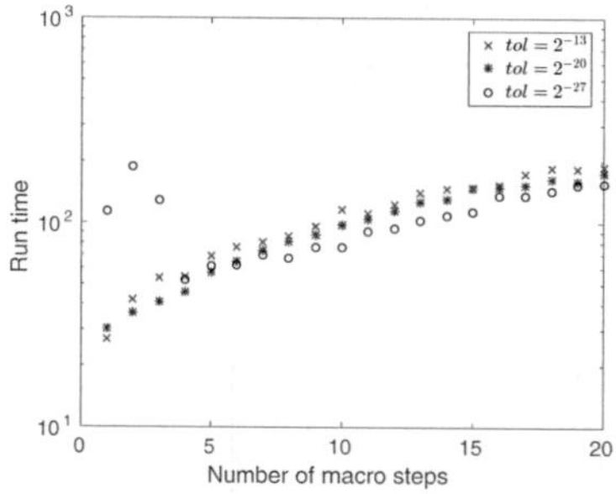

(c)

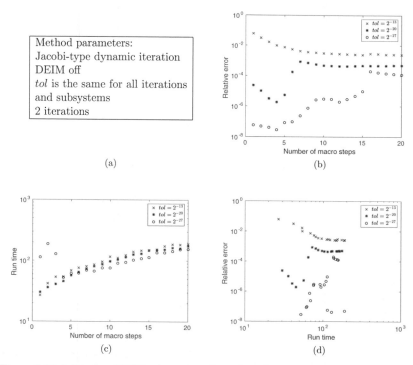

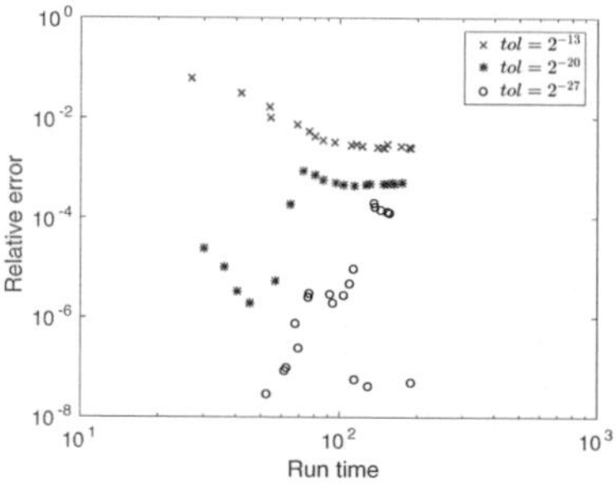

(d)

Figure 3.10: Influence of POD tolerance and number of macro steps on the relative error and run time for the Jacobi-type DIRM method

Next, we vary the number of macro steps n_T on the time interval $[0, 2]$. In dynamic iteration, increasing number of macro steps leads to smaller errors. Figure 3.10(b) basically shows this behaviour for tolerance 2^{-13} only, whereas for tolerances 2^{-20} and 2^{-27}, the relative error for $n_T = 4$ macro steps is clearly the smallest. We suspect that too small macro steps lead to bad snapshots and, therefore, to large errors. Looking at the run time in Figure 3.10(c), it increases slightly with increasing number of macro steps. This is to be expected because of the overhead computations. Only for tolerance 2^{-27} with one, two and three macro steps, the simulation is significantly longer. Finally, Figure 3.10(d) gives the correlation between the run time and the relative error. From the tested variants, four macro time steps and the POD tolerance $tol = 2^{-27}$ provides the best result.

Summarizing, we conclude that the DIRM method is very sensitive to method parameters. It is not applicable without a well-founded strategy for parameter selection. An inappropriate choice of parameters may lead to a failure of the method.

3.4 A posteriori error estimator

In this section, we present an a posteriori error estimator for the DIRM method applied to coupled ODEs. It is based on an a posteriori error estimator for DEIM derived in [WSH14].

Theorem 3.5. *Let*

$$x = \begin{bmatrix} x_1 \\ \vdots \\ x_N \end{bmatrix} \quad \text{and} \quad x^{[k]} = \begin{bmatrix} x_1^{[k]} \\ \vdots \\ x_N^{[k]} \end{bmatrix}$$

be the solutions of a coupled system (3.1.3) *and the k-th iterate of the DIRM, respectively. Then the DIRM error* $e(t) = x(t) - x^{[k]}(t)$ *is bounded on very time window* $[T_m, T_{m+1}]$ *as*

$$\|e(t)\| \leqslant \int_{T_m}^{t} \beta(s) e^{\int_s^t \alpha(\tau)\, d\tau} ds + \|e(T_m)\| e^{\int_{T_m}^t \alpha(\tau) d\tau}, \quad t \in [T_m, T_{m+1}], \tag{3.4.1}$$

where $\|e(T_0)\| = 0$, $\alpha(t) = L_2[f](x^{[k]}(t))$ *is the local logarithmic Lipschitz constant of* $f = \begin{bmatrix} f_1^T & \dots & f_N^T \end{bmatrix}^T$ *defined in* (2.4.16)*, and*

$$\beta(t) = \left\| \begin{bmatrix} f_1(x^{[k]}(t)) - f_1(\hat{x}_1^{[k]}(t)) \\ \vdots \\ f_N(x^{[k]}(t)) - f_N(\hat{x}_N^{[k]}(t)) \end{bmatrix} \right\|$$

with $\hat{x}_i^{[k]} = \left[(\hat{X}_{i,1}^{[k]})^T, \dots, (\hat{X}_{i,i-1}^{[k]})^T, (x_i^{[k]})^T, (\hat{X}_{i,i+1}^{[k]})^T, \dots, (\hat{X}_{i,N}^{[k]})^T \right]^T$ *for* $i = 1, \dots, N$.

Proof. For the sake of simplicity, we skip the time dependence in the following notation. The time derivative of the error $\|e\| = \|x - x^{[k]}\|$ can be calculated as

$$\frac{d\|e\|}{dt} = \frac{d\langle e, e\rangle^{\frac{1}{2}}}{dt} = \frac{\langle \dot{e}, e\rangle + \langle e, \dot{e}\rangle}{2\|e\|} = \frac{\langle \dot{e}, e\rangle}{\|e\|}.$$

Collecting the i-th equations from the i-th DIRM system (3.2.1), we get the system

$$\begin{aligned} \dot{x}_1^{[k]} &= f_1(\hat{x}_1^{[k]}), \\ &\vdots \\ \dot{x}_N^{[k]} &= f_N(\hat{x}_N^{[k]}). \end{aligned}$$

In order to determine the inner product $\langle \dot{e}, e\rangle$, we subtract this system from the coupled ODE system (3.1.3). Then we obtain

$$\langle \dot{e}, e\rangle = \left\langle \begin{bmatrix} f_1(x) \\ \vdots \\ f_N(x) \end{bmatrix} - \begin{bmatrix} f_1(\hat{x}_1^{[k]}) \\ \vdots \\ f_N(\hat{x}_N^{[k]}) \end{bmatrix}, e \right\rangle$$

Next, we add and substract $\left[f_1(x^{[k]})^T, \ldots, f_N(x^{[k]})^T\right]^T$ and use the Cauchy-Schwarz inequality. This leads to

$$\begin{aligned}\langle \dot{e}, e\rangle &= \left\langle \begin{bmatrix} f_1(x) \\ \vdots \\ f_N(x) \end{bmatrix} - \begin{bmatrix} f_1(x^{[k]}) \\ \vdots \\ f_N(x^{[k]}) \end{bmatrix}, e \right\rangle + \left\langle \begin{bmatrix} f_1(x^{[k]}) \\ \vdots \\ f_N(x^{[k]}) \end{bmatrix} - \begin{bmatrix} f_1(\hat{x}_1^{[k]}) \\ \vdots \\ f_N(\hat{x}_N^{[k]}) \end{bmatrix}, e \right\rangle \\ &\leqslant \left\langle \begin{bmatrix} f_1(x) \\ \vdots \\ f_N(x) \end{bmatrix} - \begin{bmatrix} f_1(x^{[k]}) \\ \vdots \\ f_N(x^{[k]}) \end{bmatrix}, e \right\rangle + \left\| \begin{bmatrix} f_1(x^{[k]}) \\ \vdots \\ f_N(x^{[k]}) \end{bmatrix} - \begin{bmatrix} f_1(\hat{x}_1^{[k]}) \\ \vdots \\ f_N(\hat{x}_N^{[k]}) \end{bmatrix} \right\| \|e\|.\end{aligned}$$

Using the local logarithmic Lipschitz constant $L_2[f](x^{[k]})$ to estimate the first inner product, we have

$$\begin{aligned}\frac{d\|e(t)\|}{dt} &= \frac{\langle \dot{e}(t), e(t)\rangle}{\|e(t)\|} \leqslant L_2[f](x^{[k]}(t))\|e(t)\| + \left\| \begin{bmatrix} f_1(x^{[k]}(t)) \\ \vdots \\ f_N(x^{[k]}(t)) \end{bmatrix} - \begin{bmatrix} f_1(\hat{x}_1^{[k]}(t)) \\ \vdots \\ f_N(\hat{x}_N^{[k]}(t)) \end{bmatrix} \right\| \\ &= \alpha(t)\|e(t)\| + \beta(t).\end{aligned}$$

Using the comparison lemma [WSH14], we obtain estimate (3.4.1). Finally, obviously $\|e(T_0)\| = \|x(t_0) - x^{[k]}(t_0)\| = 0$. □

Since the local logarithmic Lipschitz constant $L_2[f](x^{[k]}(t))$ is expensive to calculate, we suggest similarly to [WSH14] to approximate it with the logarithmic norm $L_2[J_f(x^{[k]}(t))]$ of the Jacobi matrix $J_f(x^{[k]}(t))$ of f evaluated at $x^{[k]}(t)$. In order to compute the logarithmic norm $L_2[J_f(x^{[k]}(t))] = \frac{1}{2}\lambda_{\max}\big(J_f(x^{[k]}(t)) + J_f^T(x^{[k]}(t))\big)$ we first determine a MDEIM approximation to the Jacobi matrix

$$J_f(x^{[k]}(t)) \approx \sum_{q=1}^{Q} W_q \theta_q(x^{[k]}(t))$$

as described in Section 2.4.3 and then calculate the largest eigenvalue of the symmetric matrix $M(t) = \sum_{q=1}^{Q} \theta_q(x^{[k]}(t))M_q$ with $M_q = W_q + W_q^T$ using the successive constraint method [HRSP07]. This method will be presented in the next section. The computation of $\beta(t)$ requires the evaluation of f_i at $x^{[k]}(t)$ and $\hat{x}_i^{[k]}(t)$ for $i = 1, \ldots, N$. This data can be collected through DIRM. Note that exploiting the structure of the problem can reduce the computation time for the error estimates significantly.

3.4.1 Successive constraint method

In this section, we briefly describe the successive constraint method (SCM) for computing an upper bound on

$$\lambda(\mu) = \max_{v \in \mathbb{R}^n \setminus \{0\}} \frac{v^T M(\mu) v}{v^T v}, \quad \mu \in D \subset \mathbb{R}^n \tag{3.4.2}$$

for a parameter dependent matrix $M(\mu) = \sum_{q=1}^{Q} \theta_q(\mu) M_q$ with $M_q \in \mathbb{R}^{n \times n}$ and nonlinear functions $\theta_q : D \to \mathbb{R}$. This method was first proposed in [HRSP07] for the construction of lower bounds for the coercivity and inf-sub stability constraints requested in an a posteriori error analysis of parametrized PDEs. Here, we adapt the SCM to our setting by considering a finite dimensional space $\mathbb{R}^n$ and searching for an upper bound on a discrete parameter set. Note that if $M(\mu)$ is symmetric, then by the Courant-Fischer theorem [HJ85], $\lambda(\mu)$ in (3.4.2) gives the largest eigenvalue of $M(\mu)$. We first define a set

$$Y = \left\{ y = [y_1, \ldots, y_Q]^T \in \mathbb{R}^Q \, : \, \exists v \in \mathbb{R}^n \setminus \{0\} \text{ s.t. } y_q = \frac{v^T M_q v}{v^T v},\, q = 1, \ldots, Q \right\}$$

and as objective the function $\mathcal{J}(\mu, y) = \sum_{q=1}^{Q} \theta_q(\mu) y_q$. Then (3.4.2) can be written as $\lambda(\mu) = \max_{y \in Y} \mathcal{J}(\mu, y)$. Furthermore, we introduce a bounded set

$$\begin{aligned} B_Q = & \left[\min_{v \in \mathbb{R}^n \setminus \{0\}} \frac{v^T M_1 v}{v^T v}, \max_{v \in \mathbb{R}^n \setminus \{0\}} \frac{v^T M_1 v}{v^T v} \right] \times \cdots \\ & \times \left[\min_{v \in \mathbb{R}^n \setminus \{0\}} \frac{v^T M_Q v}{v^T v}, \max_{v \in \mathbb{R}^n \setminus \{0\}} \frac{v^T M_Q v}{v^T v} \right] \subset \mathbb{R}^Q. \end{aligned}$$

For a parameter set $C_K = \{\mu_1, \ldots, \mu_K\} \subset D$ we define further the sets

$$\begin{aligned} Y_{UB}(C_K) &= \{ y \in B_Q \, : \, \mathcal{J}(\mu_k, y) \leqslant \lambda(\mu_k) \text{ for all } \mu_k \in C_K \}, \\ Y_{LB}(C_K) &= \{ \arg\max_{y \in Y} \mathcal{J}(\mu_k, y) \text{ for all } \mu_k \in C_K \} \end{aligned}$$

and bounds $\lambda_{UB}(\mu, C_K) = \max\limits_{y \in Y_{UB}(C_K)} \mathcal{J}(\mu, y)$ and $\lambda_{LB}(\mu, C_K) = \max\limits_{y \in Y_{LB}(C_K)} \mathcal{J}(\mu, y)$.

Lemma 3.6. *For a parameter set C_K, it holds*

$$\lambda_{LB}(\mu, C_K) \leqslant \lambda(\mu) \leqslant \lambda_{UB}(\mu, C_K) \text{ for all } \mu \in D. \tag{3.4.3}$$

Proof. We first observe that $Y_{LB}(C_K) \subset Y$ by definition of $Y_{LB}(C_K)$. Secondly we show that $Y \subset Y_{UB}(C_K)$. For $y = [y_1, \ldots, y_Q]^T \in Y$, we have $y_q = \frac{v^T M_q v}{v^T v}$ for some $v \neq 0$ and $q = 1, \ldots, Q$ and, therefore,

$$\min_{v \in \mathbb{R}^n \setminus \{0\}} \frac{v^T M_q v}{v^T v} \leqslant y_q \leqslant \max_{v \in \mathbb{R}^n \setminus \{0\}} \frac{v^T M_q v}{v^T v}, \qquad q = 1, \ldots, Q.$$

Thus, $y \in B_Q$. Furthermore, by the definition of $\lambda(\mu)$, we have $\mathcal{J}(\mu, y) \leqslant \lambda(\mu)$ for all $y \in Y$ and, therefore, $Y \subset Y_{UB}(C_K)$. Then (3.4.3) immediately follows from the definition of λ_{UB}, λ and λ_{LB}. □

The key idea of the SCM is to determine a parameter set C_K such that the corresponding lower and upper bounds $\lambda_{LB}(\mu, C_K)$ and $\lambda_{UB}(\mu, C_K)$ are sufficiently close to $\lambda(\mu)$. Such a set can be computed iteratively using the greedy procedure. Once the parameter set C_K is constructed, we choose the next parameter μ_{K+1} such that the error $\epsilon_K(\mu, C_K) = 1 - \frac{\lambda_{LB}(\mu, C_K)}{\lambda_{UB}(\mu, C_K)}$ is the largest in a training set D_{train} which is a discrete subset of D. Updating $C_{K+1} = C_K \cup \{\mu_{K+1}\}$, we continue the iteration until $\epsilon_K(\mu) < tol$ with a tolerance $tol \in (0, 1)$. The SCM is summarized in Algorithm 3.1. For arbitrary parameter $\mu \in D$, the upper bound $\lambda_{UB}(\mu, C_K)$ on $\lambda(\mu)$ can be determined by solving the linear program (3.4.4) with μ_j replaced by μ, which has Q variables and $2Q + K$ constraints. Thus, the complexity of this program is independent of the problem dimension n.

Algorithm 3.1: Successive constraints method

Input : $M_1, \dots, M_Q \in \mathbb{R}^{n\times n}$, $\theta(\mu) = [\theta_1(\mu), \dots, \theta_Q(\mu)]$ nonlinear function, training set $D_{train} = \{\mu_1, \dots, \mu_g\}$, maximal number of iterations $maxiter < g$, tolerance tol, number of an initial starting parameter $\mu_{i_1} \in D_{train}$

Output: a parameter set C_k

Set $C_1 = \{\mu_1\}$, $A_{UB} = []$, $b_{UB} = []$.

Compute $B_Q = \prod_{q=1}^{Q} \left[\min\limits_{v\in\mathbb{R}^n\setminus\{0\}} \dfrac{v^T M_q v}{v^T v}, \max\limits_{v\in\mathbb{R}^n\setminus\{0\}} \dfrac{v^T M_q v}{v^T v} \right]$.

for $k = 1$ *to maxiter* **do**

  Compute $\lambda_{i_k} = \max\limits_{v\in\mathbb{R}^n\setminus\{0\}} \theta(\mu_{i_k}) \left[\frac{v^T M_1 v}{v^T v}, \dots, \frac{v^T M_Q v}{v^T v}\right]^T$ and

  $v_{i_k} = \arg \max\limits_{v\in\mathbb{R}^n\setminus\{0\}} \theta(\mu_{i_k}) \left[\frac{v^T M_1 v}{v^T v}, \dots, \frac{v^T M_Q v}{v^T v}\right]^T$.

  Set $A_{UB} \leftarrow [A_{UB}, \theta^T(\mu_{i_k})]$, $b_{UB} \leftarrow [b_{UB}, \lambda_{i_k}]$.

  for $j = 1$ *to* g **do**

    Solve the linear program

$$\max \theta(\mu_j) y \quad \text{s.t.} \quad A_{UB}^T y \leqslant b_{UB}^T,\ y \in B_Q \tag{3.4.4}$$

    for y.

    Set $\lambda_{UB}(\mu_j, C_k) = \theta(\mu_j) y$.

    Compute $\lambda_{LB}(\mu_j, C_k) = \max\limits_{1\leqslant l\leqslant k} \theta(\mu_j) \left[\frac{v_{i_l}^T M_1 v_{i_l}}{v_{i_l}^T v_{i_l}}, \dots, \frac{v_{i_l}^T M_Q v_{i_l}}{v_{i_l}^T v_{i_l}}\right]^T$.

    Compute $\epsilon(\mu_j, C_k) = \frac{\lambda_{UB}(\mu_j, C_k) - \lambda_{LB}(\mu_j, C_k)}{\lambda_{UB}(\mu_j, C_k)}$.

  end

  Find $\mu_{i_{k+1}} = \arg \max\limits_{\mu_j \in D_{train}} \epsilon(\mu_j, C_k)$.

  Set $C_{k+1} = C_k \cup \{\mu_{i_{k+1}}\}$.

  if $\epsilon(\mu_{i_{k+1}}, C_k) < tol$ **then**

    break

  end

end

3.4.2 Numerical example

Here, we present some results of numerical experiments to demonstrate the effectiveness of the a posteriori DIRM error estimator (3.4.1) for the coupled system described in Section 3.3. To solve this system, we use the Jacobi-type DIRM method with $n_T = 4$ macro time steps on the time interval $[0, 2]$ and $K_m = 2$. The dimension of the reduced variables was determined using the POD tolerance $tol = 2^{-27}$. As in Section 3.3, a reference solution x_{ref} was computed by solving the semidiscretized equation (3.3) using the MATLAB function ode15s with stepsize control. To be able to calculate the absolute DIRM error $\|x_{ref}(t) - x^{[k]}(t)\|$ not only at the macro time steps, all systems of the DIRM iteration were calculated on the time grid of the reference solution x_{ref}. In Figure 3.11(a), we present the absolute DIRM error $\|x_{ref}(t) - x^{[k]}(t)\|$ and the error estimator (3.4.1) with $L_2[f](x^{[2]}(t))$ replaced by the logarithmic norm $L_2\big[J_f(x^{[2]}(t))\big]$ computed without and with MDEIM and SCM. One can see that the error estimator overestimates the true error by three orders of magnitude, but it depicts the behavior of the error very well. Even small peaks when changing from one macro time step to the other are reflected. The relative error caused by MDEIM and SCM in the error estimator is shown in Figure 3.11(b). This one is so small that it does not matter here. In this small example, the time saved by MDEIM and SCM can hardly be shown. Table 3.1 shows that a speedup can be achieved with MDEIM and SCM if the generated MDEIM basis can be used long enough. In our experiments, the reduced basis was only created in the first macro time step and then reused in other macro time steps. Accordingly, the run time in the first time window $[T_0, T_1]$ is significantly longer than in the other macro time steps.

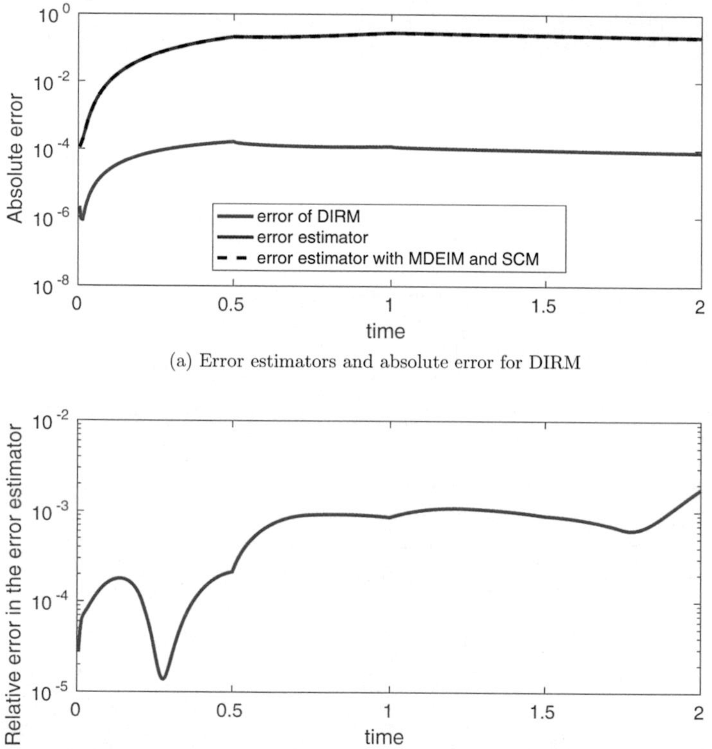

(a) Error estimators and absolute error for DIRM

(b) Relative error between error estimators without and with MDEIM and SCM

Figure 3.11: Error estimation without and with using MDEIM and SCM

	without MDEIM and SCM	with MDEIM and SCM
1. macro time step	4.68	7.12
2. macro time step	4.65	3.45
3. macro time step	4.83	3.46
4. macro time step	4.71	1.93

Table 3.1: Run time in seconds for the error estimator without and with MDEIM and SCM

4 Model reduction for magneto-quasistatic problems

Nowadays, integrated circuits play an increasingly important role. Modelling of electromagnetic effects in high-frequency and high-speed electronic systems leads to coupled electromagnetic-circuit models of high complexity. The development of efficient, fast and accurate simulation tools for such models is of great importance in the computer-aided design of electromagnetic structures offering significant savings in production cost and time. In this chapter, we consider model order reduction of magneto-quasistatic (MQS) models obtained from Maxwell's equations by assuming that the contribution of displacement current is negligible compared to the conductive currents. MQS equations are used for modeling of low-frequency electromagnetic devices like transformers, induction sensors and generators. Due to the presence of non-conducting subdomains such equations form a system of partial differential-algebraic equations (PDAEs), whose dynamics are restricted to a manifold described by algebraic constraints. A spatial discretization of MQS problems using the finite integration technique (FIT) [Wei77] or the FEM [Néd80, Bos98, Mon03] leads to DAE systems which are singular in the 3D case. Here, we consider only the FEM discretization. The algebraic constraints should be treated carefully when approximating the system. A naive application of the existing model reduction methods to DAEs may lead to an inaccurate approximation and physically meaningless results [GSW13, KS15, Sty11]. We will exploit the special block structure of the semidiscretized MQS system to transform it into a system of ODEs. Unfortunately, the transformation to the ODE form requires the computation of an orthonormal basis of a certain subspace and destroys the sparsity of the system matrices. To overcome these computational difficulties and to construct efficient model reduction methods for linear and nonlinear MQS systems, we will again use the underlying structure of the problem.

4.1 Maxwell's equations

First, we introduce a physical model. The dynamical behavior of the electromagnetic field is described by Maxwell's equations in a differential form

$$\nabla \times \mathbf{H} = \mathbf{J}_c + \mathbf{J}_s + \frac{\partial \mathbf{D}}{\partial t}, \tag{4.1.1a}$$

$$\nabla \times \mathbf{E} = -\frac{\partial \mathbf{B}}{\partial t}, \tag{4.1.1b}$$

$$\nabla \cdot \mathbf{B} = 0, \tag{4.1.1c}$$

$$\nabla \cdot \mathbf{D} = \rho, \tag{4.1.1d}$$

with the physical quantities:

- the magnetic field intensity $\mathbf{H}$,
- the magnetic flux density $\mathbf{B}$,
- the electric field intensity $\mathbf{E}$,
- the electric flux density $\mathbf{D}$,
- the conduction current density $\mathbf{J}_c$,
- the external source current density $\mathbf{J}_s$,
- the electric charge density ρ,

which depend on the spatial variable $\xi \in \mathbb{R}^3$ and time $t \in [0, T] \subset \mathbb{R}$, see [Max65]. The first six quantities are $\mathbb{R}^3$-valued functions, whereas ρ is a scalar function. Additionally, there are the material equations

$$\mathbf{B} = \boldsymbol{\mu}\mathbf{H}, \tag{4.1.2a}$$

$$\mathbf{D} = \boldsymbol{\epsilon}\mathbf{E}, \tag{4.1.2b}$$

$$\mathbf{J}_c = \boldsymbol{\sigma}\mathbf{E} \tag{4.1.2c}$$

with the material parameters:

- the magnetic permeability $\boldsymbol{\mu}$,
- the electric permitivity $\boldsymbol{\epsilon}$,
- the electric conductivity $\boldsymbol{\sigma}$.

We consider only isotropic materials without hysteresis effects. In this case, the material quantities are scalar functions, where $\boldsymbol{\epsilon}$ and $\boldsymbol{\sigma}$ depend only on the space variable, and the permeability can be represented as a function of the magnetic field $\mathbf{H}$ such that

$$\mathbf{B} = \boldsymbol{\mu}(\cdot, \|\mathbf{H}\|)\mathbf{H}. \tag{4.1.3}$$

Furthermore, we introduce the magnetic reluctivity $\boldsymbol{\nu}$ satisfying $\boldsymbol{\nu}(\cdot, \|\mathbf{B}\|) = \frac{1}{\boldsymbol{\mu}(\cdot, \|\mathbf{H}\|)}$ which lead to the relation

$$\mathbf{H} = \boldsymbol{\nu}(\cdot, \|\mathbf{B}\|)\mathbf{B}. \tag{4.1.4}$$

For derivation of Maxwell's equations and detailed description of materials, we refer to [Gri13, Jac99, Pec04].

Due to the divergence-free property for the magnetic flux density $\mathbf{B}$, one can find a magnetic vector potential $\mathbf{A}$ and an electric scalar potential φ satisfying

$$\mathbf{B} = \nabla \times \mathbf{A}, \tag{4.1.5}$$

$$\mathbf{E} = -\frac{\partial \mathbf{A}}{\partial t} - \nabla \varphi. \tag{4.1.6}$$

Note that the potentials $\mathbf{A}$ and φ are not unique, whereas $\mathbf{B}$ and $\mathbf{E}$ are uniquely defined. Consider the gauge transformation

$$\tilde{\mathbf{A}} = \mathbf{A} + \nabla\psi, \quad \tilde{\varphi} = \varphi - \frac{\partial \psi}{\partial t} \tag{4.1.7}$$

with an arbitrary scalar function ψ. One can show that $\mathbf{B}$ and $\mathbf{E}$ are gauge invariant in the sense that they remain unchanged under the gauge transformation [Bos98, Jac02]. A uniqueness of $\mathbf{A}$ and φ up to a constant scalar field can be archived by the Coulomb gauge

$$\nabla \cdot \mathbf{A} = 0 \tag{4.1.8}$$

or the Lorenz gauge

$$\epsilon\mu\frac{\partial \varphi}{\partial t} + \nabla \cdot \mathbf{A} = 0.$$

For further discussions on gauge, we refer to [Bau12, Bos98, Jac02, Sch11].

4.2 Magneto-quasistatic model

MQS equations can be considered as an approximation of Maxwell's equations (4.1.1), (4.1.3) for low-frequency problems [Dir96, HM89]. We assume that the displacement currents are negligible in comparison to the conductive currents, i.e.,

$$\frac{\partial \mathbf{D}}{\partial t} = 0. \tag{4.2.1}$$

Inserting the material equations (4.1.2c) and (4.1.4) into equation (4.1.1a) and using the condition (4.2.1) and the potential equations (4.1.5) and (4.1.6), we obtain the magnetic vector potential formulation

$$\boldsymbol{\sigma}\frac{\partial \mathbf{A}}{\partial t} + \nabla \times (\boldsymbol{\nu}(\cdot, \|\nabla \times \mathbf{A}\|)\nabla \times \mathbf{A}) = -\boldsymbol{\sigma}\nabla\varphi + \mathbf{J}_s. \tag{4.2.2}$$

Boundary and initial conditions Maxwell's equations are, in general, defined on an infinite domain. We cannot numerically capture this and, therefore, consider the equations on a bounded simply connected polyhedral region $\Omega \subset \mathbb{R}^3$. There is three types of boundary conditions:

1. Dirichlet boundary condition

$$\mathbf{A} \times n_0 = \mathbf{A}_D \qquad \text{on } \partial\Omega \times (0, T)$$

with an outer unit normal n_0 to the boundary $\partial\Omega$ and a given function $\mathbf{A}_D$,

2. Neumann boundary condition

$$\boldsymbol{\nu}(\nabla \times \mathbf{A}) \times n_0 = \mathbf{H}_N \qquad \text{on } \partial\Omega \times (0, T)$$

with an outer unit normal n_0 to the boundary $\partial\Omega$ and a given function $\mathbf{H}_N$,

3. anti-periodic boundary condition

$$\big(\nabla \times \mathbf{A}(\xi,t)\big) \cdot n_0^+ = -\big(\nabla \times \mathbf{A}(s(\xi),t)\big) \cdot n_0^- \qquad \text{for } (\xi,t) \in \Gamma_+ \times (0,T)$$

with two parts of the boundary Γ_+ and Γ_- such that $\partial\Omega = \Gamma_+ \cup \Gamma_-$, the corresponding outer unit normals n_0^+ and n_0^- and a mapping $s : \Gamma_+ \to \Gamma_-$.

For the unique solvability of the problem, at least a part of the boundary must be provided with a boundary condition of the Dirichlet type, see [QV94]. A description of the effects of different boundary conditions can be found in [Sch11]. We use a homogeneous Dirichlet boundary condition

$$\mathbf{A} \times n_0 = 0 \qquad \text{on } \partial\Omega \times (0,T).$$

In addition, we specify the initial condition

$$\mathbf{A}(\cdot,0) = \mathbf{A}_0 \qquad \text{in } \Omega.$$

Material parameters Let $\Omega \subset \mathbb{R}^3$ be a bounded domain, with a conducting subdomain Ω_1 and a non-conducting subdomain Ω_2 such that $\Omega_1 \cap \Omega_2 = \emptyset$ and $\overline{\Omega}_1 \cup \overline{\Omega}_2 = \overline{\Omega}$. We assume that the material parameters $\boldsymbol{\sigma}$ and $\boldsymbol{\nu}$ are different in the conducting and non-conducting subdomains. We consider

- the electric conductivity

$$\boldsymbol{\sigma}(\xi) = \begin{cases} \boldsymbol{\sigma}_1 & \text{in } \Omega_1, \\ 0 & \text{in } \Omega_2, \end{cases}$$

  with a constant $\boldsymbol{\sigma}_1 > 0$ and

- the magnetic reluctivity

$$\boldsymbol{\nu}(\xi,\varrho) = \begin{cases} \boldsymbol{\nu}_1(\varrho) & \text{in } \Omega_1, \\ \boldsymbol{\nu}_2 & \text{in } \Omega_2, \end{cases} \tag{4.2.3}$$

  with a constant $\boldsymbol{\nu}_2 > 0$ and a function $\boldsymbol{\nu}_1 : \mathbb{R}_0^+ \to \mathbb{R}_0^+$ satisfying the following conditions:

  1. $\boldsymbol{\nu}_1$ is continuously differentiable;
  2. $\boldsymbol{\nu}_1(\cdot)\cdot$ is strongly monotone with monotonicity constant $m_{\nu,1} > 0$, i.e.,

$$(\boldsymbol{\nu}_1(\varrho)\varrho - \boldsymbol{\nu}_1(\hat{\varrho})\hat{\varrho})(\varrho - \hat{\varrho}) \geqslant m_{\nu,1}(\varrho - \hat{\varrho})^2 \quad \text{for all } \varrho, \hat{\varrho} \geqslant 0; \tag{4.2.4}$$

  3. $\boldsymbol{\nu}_1(\cdot)\cdot$ is Lipschitz continuous with Lipschitz constant $L_{\nu,1} > 0$, i.e.,

$$|\boldsymbol{\nu}_1(\varrho)\varrho - \boldsymbol{\nu}_1(\hat{\varrho})\hat{\varrho}| \leqslant L_{\nu,1}|\varrho - \hat{\varrho}| \quad \text{for all } \varrho, \hat{\varrho} \geqslant 0. \tag{4.2.5}$$

Obviously, the function $\boldsymbol{\nu}$ inherits the properties of $\boldsymbol{\nu}_1$. Conditions 2 and 3 imply that for all $\xi \in \Omega$, $\boldsymbol{\nu}(\xi,\cdot)\cdot$ is strongly monotone with monotonicity constant $m_\nu = \min(m_{\nu,1}, \boldsymbol{\nu}_2)$, and it is Lipschitz continuous with Lipschitz constant $L_\nu = \max(L_{\nu,1}, \boldsymbol{\nu}_2)$. Moreover, it follows from conditions 1 and 2 that

$$\boldsymbol{\nu}(\xi, \varrho) \geqslant m_\nu \quad \text{for all } (\xi, \varrho) \in \Omega \times \mathbb{R}_0^+. \tag{4.2.6}$$

Indeed, taking $\hat{\varrho} = 0$ in (4.2.4), we obtain $\boldsymbol{\nu}_1(\varrho)\varrho^2 \geqslant m_{\nu,1}\varrho^2$ and, therefore, $\boldsymbol{\nu}_1(\varrho) \geqslant m_{\nu,1}$ for all $\varrho \in \mathbb{R}^+$. This inequality is also valid for $\varrho = 0$, because of the continuity of $\boldsymbol{\nu}_1$. Thus, (4.2.6) is fulfilled.

Note that due to (4.1.7), we can choose $\mathbf{A}$ such that $\boldsymbol{\sigma}\nabla\varphi = 0$. Then (4.2.2) simplifies to the equation

$$\boldsymbol{\sigma}\frac{\partial \mathbf{A}}{\partial t} + \nabla \times (\boldsymbol{\nu}(\cdot, \|\nabla \times \mathbf{A}\|)\nabla \times \mathbf{A}) = \mathbf{J}_s \tag{4.2.7}$$

guaranteeing , $\mathbf{A}$ is uniquely determined on Ω_1. To get the uniqueness of Ω_1 on Ω_2, we use the Coulomb gauge (4.1.8) on Ω_2, see [Sch11].

Stranded conductors The coupling of electromagnetic devices to a circuit is established by the voltage drop v and the electric current ι through the electromagnetic conductive contacts. This can be realized as a solid conductor model, where solid conductors behave as voltage-driven circuit elements, or as a stranded conductor model, where stranded conductors behave as current-driven circuit elements, see [Ben07, SGW13]. Since both conductor models are equivalent in the sense that the solid conductor can be written as a stranded conductor with a particular conductivity matrix [Sch11], we restrict ourselves to the stranded conductor model. A stranded conductor consists of several thin wires (strands) of small cross-sectional area forming a coil. For such a conductor, it is assumed that the current density is homogeneously distributed. For a stranded conductor model with m terminals, the source current density is given by

$$\mathbf{J}_s = \boldsymbol{\chi}_{\text{str}}\iota, \tag{4.2.8}$$

where a winding function $\boldsymbol{\chi}_{\text{str}} : \Omega \to \mathbb{R}^{3\times m}$ satisfies the following conditions:

$$\boldsymbol{\chi}_{\text{str}} \text{ is divergence-free, i.e., } \nabla \cdot \boldsymbol{\chi}_{\text{str}} = 0 \quad \text{in } \Omega, \tag{4.2.9a}$$
$$\operatorname{supp}((\boldsymbol{\chi}_{\text{str}})_i) \cap \Omega_2 \neq \emptyset \text{ for } i = 1, \ldots, m, \tag{4.2.9b}$$
$$\operatorname{supp}((\boldsymbol{\chi}_{\text{str}})_i) \cap \operatorname{supp}((\boldsymbol{\chi}_{\text{str}})_j) = \emptyset \text{ for } i, j = 1, \ldots, m,\ i \neq j, \tag{4.2.9c}$$

where $(\boldsymbol{\chi}_{\text{str}})_i$ denotes the i-th column of $\boldsymbol{\chi}_{\text{str}}$. Condition (4.2.9b) means that the coupling is not only on the conductive part, whereas condition (4.2.9c) implies that the terminals do not overlap. Using Ohm's law for the resistive voltage drop and Faraday's law for the induced voltage drop, one can compute the complete voltage drop as

$$v = \mathcal{R}\iota + \frac{\partial}{\partial t}\int_\Omega \boldsymbol{\chi}_{\text{str}}^T \mathbf{A}\, d\xi,$$

where $\mathcal{R}$ describes the resistance of the winding. It is given by

$$\mathcal{R} = \int_\Omega \frac{\chi_{\rm str}^T \chi_{\rm str}}{\boldsymbol{\sigma}_{coil}\gamma}\, d\xi,$$

where $\boldsymbol{\sigma}_{coil}$ is a electric conductivity of the coil and $\gamma \in (0,1]$ is a filling factor.

Summarizing, we obtain the MQS model

$$\begin{aligned}
\boldsymbol{\sigma}\frac{\partial \mathbf{A}}{\partial t} + \nabla \times (\boldsymbol{\nu}(\cdot, \|\nabla \times \mathbf{A}\|)\nabla \times \mathbf{A}) &= \chi_{\rm str}\iota \quad && \text{in } \Omega \times (0,T), && (4.2.10a)\\
\frac{\partial}{\partial t}\int_\Omega \chi_{\rm str}^T \mathbf{A}\, d\xi + \mathcal{R}\iota &= v && \text{in } (0,T), && (4.2.10b)\\
\mathbf{A} \times n_0 &= 0 && \text{on } \partial\Omega \times (0,T), && (4.2.10c)\\
\mathbf{A}(\cdot, 0) &= \mathbf{A}_0 && \text{in } \Omega. && (4.2.10d)
\end{aligned}$$

This model can be considered as a control system of partial-integro-differential-algebraic equations with the input v, the state $[\mathbf{A}^T, \iota^T]^T$ and the output ι. Such a system has been studied in [Alt13, MHCG17, Sch11] in the context of coupled field-circuit problems. Our goal is now to develop efficient model reduction methods for (4.2.10) which exploit the underlying structure of the problem.

4.3 2D Magneto-quasistatic problems

First, we consider 2D MQS systems in more details. We discuss a spatial discretization of such systems using FEM which leads to DAEs and also study the index and passivity of the semi-discretized system. In Sections 4.4 and 4.5, we present passivity-preserving model reduction methods for linear and nonlinear 2D MQS equations.

Let $\Omega \subset \mathbb{R}^2$ be a bounded connected domain with a Lipschitz continuous boundary $\partial\Omega$. For the MQS problem on such a domain Ω, we assume that the magnetic field $\mathbf{H}$ lies in the (ξ_1, ξ_2)-plane meaning that it is a three-dimensional vector field which does not depend on the ξ_3-variable, i.e.,

$$\mathbf{H} = \begin{bmatrix} \mathbf{H}_1(\xi_1, \xi_2) \\ \mathbf{H}_2(\xi_1, \xi_2) \\ 0 \end{bmatrix}.$$

Since $\mathbf{H}$ and $\mathbf{B}$ are related via the magnetic permeability as in (4.1.3), $\mathbf{B}$ has a similar form

$$\mathbf{B} = \begin{bmatrix} \mathbf{B}_1(\xi_1, \xi_2) \\ \mathbf{B}_2(\xi_1, \xi_2) \\ 0 \end{bmatrix}.$$

Furthermore, equations (4.2.7) and (4.2.8) give directly that

$$\mathbf{J}_s = \begin{bmatrix} 0 \\ 0 \\ \mathbf{J}_{s3}(\xi_1, \xi_2) \end{bmatrix}, \qquad \chi_{\rm str} = \begin{bmatrix} 0 \\ 0 \\ \chi_{{\rm str},3}(\xi_1, \xi_2) \end{bmatrix}.$$

It follows from the equation $\mathbf{B} = \nabla \times \mathbf{A}$ that

$$\mathbf{A} = \begin{bmatrix} 0 \\ 0 \\ \mathbf{A}_3(\xi_1, \xi_2) \end{bmatrix}. \tag{4.3.1}$$

This implies

$$\|\nabla \times \mathbf{A}\| = \sqrt{\left(\frac{\partial \mathbf{A}_3}{\partial \xi_1}\right)^2 + \left(\frac{\partial \mathbf{A}_3}{\partial \xi_2}\right)^2} = \|\nabla \mathbf{A}_3\|$$

and

$$\nabla \times (\boldsymbol{\nu}(\cdot, \|\nabla \times \mathbf{A}\|)\nabla \times \mathbf{A}) = \begin{bmatrix} 0 \\ 0 \\ -\nabla \cdot (\boldsymbol{\nu}(\cdot, \|\nabla \mathbf{A}_3\|)\nabla \mathbf{A}_3) \end{bmatrix}.$$

Thus, equation (4.2.10a) can be simplified to

$$\boldsymbol{\sigma}\frac{\partial}{\partial t}\mathbf{A}_3 - \nabla \cdot \left(\boldsymbol{\nu}\left(\cdot, \|\nabla \mathbf{A}_3\|\right) \nabla \mathbf{A}_3\right) = \boldsymbol{\chi}_{\mathrm{str},3}\iota$$

and equation (4.2.10b) to

$$\frac{\partial}{\partial t}\int_\Omega \boldsymbol{\chi}_{\mathrm{str},3}^T \mathbf{A}_3 \, d\xi + \mathcal{R}\iota = v.$$

The boundary condition (4.2.10c) with $\mathbf{A}$ as in (4.3.1) and the normal vector $n_0 = \left[n_{01},\, n_{02},\, 0\right]^T$ takes the form

$$0 = \begin{bmatrix} 0 \\ 0 \\ \mathbf{A}_3 \end{bmatrix} \times \begin{bmatrix} n_{01} \\ n_{02} \\ 0 \end{bmatrix} = \begin{bmatrix} -\mathbf{A}_3 n_{02} \\ \mathbf{A}_3 n_{01} \\ 0 \end{bmatrix}.$$

Since the outer normal n_0 is not equal to zero, this condition is equivalent to $\mathbf{A}_3 = 0$ on $\partial\Omega \times (0, T)$. Introducing a new variable $\phi = \mathbf{A}_3$, we get the 2D MQS system

$$\boldsymbol{\sigma}\frac{\partial \phi}{\partial t} - \nabla \cdot (\boldsymbol{\nu}(\cdot, \|\nabla \phi\|)\nabla \phi) = \boldsymbol{\chi}_{\mathrm{str},3}\iota \qquad \text{in } \Omega \times (0, T), \tag{4.3.2a}$$

$$\frac{\partial}{\partial t}\int_\Omega \boldsymbol{\chi}_{\mathrm{str},3}^T \phi \, d\xi + \mathcal{R}\iota = v \qquad \text{in } (0, T), \tag{4.3.2b}$$

$$\phi = 0 \qquad \text{on } \partial\Omega \times (0, T), \tag{4.3.2c}$$

$$\phi(\cdot, 0) = \phi_0 \qquad \text{in } \Omega. \tag{4.3.2d}$$

4.3.1 Weak formulation

We now derive a weak formulation for the 2D MQS system (4.3.2). For this purpose, we multiply equation (4.3.2a) with a test function $\psi \in H_0^1(\Omega)$ and integrate the resulting equation over the domain Ω. Using Green's formula and the boundary condition (4.3.2c), we obtain the variational equation

$$\int_\Omega \boldsymbol{\sigma}\frac{\partial \phi}{\partial t}\psi \, d\xi + \int_\Omega \boldsymbol{\nu}(\cdot, \|\nabla \phi\|)\nabla \phi \cdot \nabla \psi \, d\xi = \int_\Omega \psi \boldsymbol{\chi}_{\mathrm{str},3}\iota \, d\xi. \tag{4.3.3}$$

We are searching for a weak solution $\phi \in \mathcal{L}^2(0,T;H_0^1(\Omega))$ satisfying (4.3.3), (4.3.2b) and (4.3.2d) for all test functions $\psi \in H_0^1(\Omega)$, where $\phi_0 \in \mathcal{L}^2(\Omega)$. The existence and uniqueness of the solution for such a problem has been investigated in [JRS14]. If there is no conductive material present or temporal changes are very slow, then one gets the magnetostatic system studied in [Pec04]. The MQS problem with piece-wise constant reluctivity on the domain Ω has been considered in [NT13].

We present now some useful results for the MQS system in the weak formulation. To this end, we introduce the functional $b : H_0^1(\Omega) \times H_0^1(\Omega) \to \mathbb{R}$ as

$$b(\varphi, \psi) = \int_\Omega \boldsymbol{\nu}(\cdot, \|\nabla\varphi\|)\nabla\varphi \cdot \nabla\psi \, d\xi. \tag{4.3.4}$$

Clearly, b is a semilinear form, i.e., it is linear in the second argument. In the following theorems, the Lipschitz continuity and strongly monotonicity of this form are shown.

Theorem 4.1. *Consider a semilinear form $b : H_0^1(\Omega)\times H_0^1(\Omega) \to \mathbb{R}$ given in (4.3.4), where $\boldsymbol{\nu}(\xi, \cdot)\cdot$ is Lipschitz continuous for all $\xi \in \Omega$ with Lipschitz constant $L_{\boldsymbol{\nu}}$. Then b is Lipschitz continuous in the first argument with Lipschitz constant $3L_{\boldsymbol{\nu}}$, i.e., it holds*

$$|b(\varphi_1, \psi) - b(\varphi_2, \psi)| \leqslant 3L_{\boldsymbol{\nu}}\|\varphi_1 - \varphi_2\|_{H_0^1(\Omega)}\|\psi\|_{H_0^1(\Omega)}$$

for all $\varphi_1, \varphi_2, \psi \in H_0^1(\Omega)$.

Proof. The proof can be found in [Pec04, Lemma 2.9]. □

Theorem 4.2. *Consider a semilinear form $b : H_0^1(\Omega)\times H_0^1(\Omega) \to \mathbb{R}$ given in (4.3.4), where $\boldsymbol{\nu}(\cdot)\cdot$ is strongly monoton for all $\xi \in \Omega$ with monotonicity constant $m_{\boldsymbol{\nu}}$. Then b satisfies*

$$b(\varphi, \varphi - \psi) - b(\psi, \varphi - \psi) \geqslant m_{\boldsymbol{\nu}}\|\nabla\varphi - \nabla\psi\|^2_{\mathcal{L}^2(\Omega)}$$

for all $\varphi, \psi \in H_0^1(\Omega)$.

Proof. See [Pec04, Lemma 2.8]. □

4.3.2 Discretization

For a spatial discretization of the MQS problem (4.3.2), we use the Lagrange FEM introduced in Section 2.5.2. Let $\{\psi_i\}_{i=1}^{n_a}$ be a basis of the finite-dimensional trial space $\mathcal{V}_h \subset H_0^1(\Omega)$ and let the finite-dimensional test space be chosen to be equal to $\mathcal{V}_h$. Substituting an approximation

$$\phi(\xi, t) \approx \sum_{i=1}^{n_a} \alpha_i(t)\psi_i(\xi)$$

into (4.3.3) and taking the basis functions ψ_j as the test functions, we obtain a nonlinear DAE control system

$$\mathcal{E}\dot{x} = \mathcal{A}(x)x + \mathcal{B}u, \tag{4.3.5a}$$

$$y = \mathcal{C}x \tag{4.3.5b}$$

with the state vector $x = \left[a^T,\ \iota^T\right]^T$, the control $u = v$, the output $y = \iota$, and

$$\mathcal{E} = \begin{bmatrix} \mathcal{M} & 0 \\ \mathcal{X}^T & 0 \end{bmatrix}, \qquad \mathcal{A}(x) = \begin{bmatrix} -\mathcal{K}(a) & \mathcal{X} \\ 0 & -\mathcal{R} \end{bmatrix}, \qquad \mathcal{B} = \mathcal{C}^T = \begin{bmatrix} 0 \\ I_m \end{bmatrix}. \tag{4.3.6}$$

Here, $a = [\alpha_1, \dots, \alpha_{n_a}]^T$ is a semidiscretized vector of magnetic potentials. Furthermore, $\mathcal{M}$ is a conductivity matrix, $\mathcal{K}(a)$ is a discrete curl-curl matrix, and $\mathcal{X}$ is a coupling matrix with the entries

$$\mathcal{M}_{i,j} = \int_\Omega \sigma \psi_j \psi_i \, d\xi, \qquad i,j = 1,\dots,n_a, \tag{4.3.7}$$

$$\mathcal{K}_{i,j}(a) = \int_\Omega \nu(\cdot, \|\sum_{k=1}^{n_a} \alpha_k \nabla\psi_k\|) \nabla\psi_j \cdot \nabla\psi_i \, d\xi, \quad i,j = 1,\dots,n_a, \tag{4.3.8}$$

$$\mathcal{X}_{i,j} = \int_\Omega \chi_{\mathrm{str},3,j} \psi_i \, d\xi, \qquad i = 1,\dots,n_a,\ j = 1,\dots,m, \tag{4.3.9}$$

respectively, where $\chi_{\mathrm{str},3,j}$ denotes the j-th column of $\chi_{\mathrm{str},3}$. Clearly, $\mathcal{M}$ and $\mathcal{K}(a)$ are both symmetric.

We now show that the matrix $\mathcal{K}(a)$ is positive definite for all a and the nonlinear function $\mathcal{K}(a)a$ is strongly monotone. For this purpose, we define a symmetric matrix K_1 with the components

$$\left(K_1\right)_{ij} = \int_\Omega \nabla\psi_j \cdot \nabla\psi_i \, d\xi, \quad i,j = 1,\dots,n_a, \tag{4.3.10}$$

and transfer the statement of Theorem 4.2 in the infinite-dimensional case to the finite-dimensional function $\mathcal{K}(a)a$.

Theorem 4.3. *Consider a nonlinear function $g : \mathbb{R}^{n_a} \to \mathbb{R}^{n_a}$ given by $g(a) = \mathcal{K}(a)a$ for all $a \in \mathbb{R}^{n_a}$, where $\mathcal{K}(a)$ is as in (4.3.8). If $\nu(\xi, \cdot)\cdot$ is strongly monotone for all $\xi \in \Omega$ with monotonicity constant m_ν, then g is strongly monotone with monotonicity constant $m_g = m_\nu \lambda_{\min}(K_1) > 0$, where $\lambda_{\min}(K_1)$ denotes the smallest eigenvalue of the matrix K_1.*

Proof. First, observe that by Theorem 2.33 the matrix K_1 is positive definite. For all $a = [\alpha_1, \dots, \alpha_{n_a}]^T$ and $w = [w_1, \dots, w_{n_a}]^T$, we obtain using Theorem 4.2 that

$$\begin{aligned}
\langle g(a) - g(w), a - w\rangle &= \langle \mathcal{K}(a)a - \mathcal{K}(w)w, a - w\rangle \\
&= b\Big(\sum_{j=1}^{n_a} \alpha_j\psi_j, \sum_{j=1}^{n_a} \alpha_j\psi_j - \sum_{j=1}^{n_a} w_j\psi_j\Big) \\
&\quad - b\Big(\sum_{j=1}^{n_a} w_j\psi_j, \sum_{j=1}^{n_a} \alpha_j\psi_j - \sum_{j=1}^{n_a} w_j\psi_j\Big) \\
&\geqslant m_\nu \Big\|\sum_{j=1}^{n_a} (\alpha_j - w_j)\nabla\psi_j\Big\|^2_{\mathcal{L}^2(\Omega)} \\
&= m_\nu (a - w)^T K_1 (a - w) \\
&\geqslant m_\nu \lambda_{\min}(K_1)\|a - w\|^2.
\end{aligned}$$

The last inequality follows from the Courant-Fischer theorem [HJ85, Theorem 4.2.6]. Thus, g is strongly monotone with the monotonicity constant $m_g = m_\nu \lambda_{\min}(K_1)$. □

The following theorem shows additionally that the matrix $\mathcal{K}(\boldsymbol{a})$ is positive definite for all $\boldsymbol{a}$.

Theorem 4.4. *Let $\mathcal{K}(\boldsymbol{a})$ be as in (4.3.8) and let $\boldsymbol{\nu}$ be as in (4.2.3), where $\boldsymbol{\nu}_1$ is continuous and $\boldsymbol{\nu}_1(\cdot)\cdot$ is strongly monotone. Then $\mathcal{K}(\boldsymbol{a})$ is positive definite for all $\boldsymbol{a} \in \mathbb{R}^{n_a}$.*

Proof. For all $w \in \mathbb{R}^{n_a} \setminus \{0\}$ and all $\boldsymbol{a} \in \mathbb{R}^{n_a}$, we have using (4.2.6) that

$$
\begin{aligned}
w^T \mathcal{K}(\boldsymbol{a}) w &= \int_\Omega \boldsymbol{\nu}(\cdot, \|\sum_{k=1}^{n_a} \alpha_k \nabla \psi_k\|) \sum_{j=1}^{n_a} w_j \nabla \psi_j \cdot \sum_{i=1}^{n_a} w_i \nabla \psi_i \, d\xi \\
&\geqslant m_\nu \int_\Omega \sum_{j=1}^{n_a} w_j \nabla \psi_j \cdot \sum_{i=1}^{n_a} w_i \nabla \psi_i \, d\xi \\
&= m_\nu \|\sum_{i=1}^{n_a} w_i \nabla \psi_i\|^2_{\mathcal{L}^2(\Omega)} \\
&\geqslant m_\nu \lambda_{\min}(K_1) \|w\|^2 > 0.
\end{aligned}
$$

Thus, $\mathcal{K}(\boldsymbol{a})$ is positive definite. □

Reordering the state variables accordingly to the conducting and non-conducting subdomains Ω_1 and Ω_2, we can partition the vector $\boldsymbol{a}$ and the matrices $\mathcal{M}$, $\mathcal{K}(\boldsymbol{a})$ and $\mathcal{X}$ as

$$
\boldsymbol{a} = \begin{bmatrix} a_1 \\ a_2 \end{bmatrix}, \quad \mathcal{M} = \begin{bmatrix} \mathcal{M}_{11} & 0 \\ 0 & 0 \end{bmatrix}, \quad \mathcal{K}(\boldsymbol{a}) = \begin{bmatrix} \mathcal{K}_{11}(a_1) & \mathcal{K}_{12} \\ \mathcal{K}_{21} & \mathcal{K}_{22} \end{bmatrix}, \quad \mathcal{X} = \begin{bmatrix} \mathcal{X}_1 \\ \mathcal{X}_2 \end{bmatrix}, \tag{4.3.11}
$$

where $a_1 \in \mathbb{R}^{n_1}$, $a_2 \in \mathbb{R}^{n_2}$, $\mathcal{M}_{11} \in \mathbb{R}^{n_1 \times n_1}$ is symmetric and positive definite, the matrices $\mathcal{K}_{11} \in \mathbb{R}^{n_1 \times n_1}$, $\mathcal{K}_{21} = \mathcal{K}_{12}^T \in \mathbb{R}^{n_2 \times n_1}$ and $\mathcal{K}_{22} \in \mathbb{R}^{n_2 \times n_2}$ are constant, $\mathcal{X}_1 \in \mathbb{R}^{n_1 \times m}$, and $\mathcal{X}_2 \in \mathbb{R}^{n_2 \times m}$. Note that nonlinearity in $\mathcal{K}(\boldsymbol{a})$ occurs only in the block $\mathcal{K}_{11}$ since the magnetic reluctivity $\boldsymbol{\nu}$ is constant on the subdomain Ω_2. The conditions (4.2.9b) and (4.2.9c) on the winding function $\boldsymbol{\chi}_{\text{str}}$ imply that the coupling matrix $\mathcal{X}$ and the block $\mathcal{X}_2$ have both full column rank. Taking into account the definition of the magnetic reluctivity $\boldsymbol{\nu}$ in (4.2.3), the matrix $\mathcal{K}(\boldsymbol{a})$ in (4.3.11) can be decomposed into the nonlinear and linear parts as

$$
\mathcal{K}(\boldsymbol{a}) = \begin{bmatrix} \mathcal{K}_{11,n}(a_1) & 0 \\ 0 & 0 \end{bmatrix} + \begin{bmatrix} \mathcal{K}_{11,l} & \mathcal{K}_{12} \\ \mathcal{K}_{21} & \mathcal{K}_{22} \end{bmatrix}, \tag{4.3.12}
$$

where the entries of $\mathcal{K}_{11,n}(a_1)$ are given by

$$
(\mathcal{K}_{11,n}(a_1))_{i,j} = \int_{\Omega_1} \boldsymbol{\nu}_1(\|\sum_{k=1}^{n_1} \alpha_k \nabla \psi_k\|) \nabla \psi_j \cdot \nabla \psi_i \, d\xi, \quad i,j = 1, \ldots, n_1, \tag{4.3.13}
$$

and the entries of

$$
\mathcal{K}_l = \begin{bmatrix} \mathcal{K}_{11,l} & \mathcal{K}_{12} \\ \mathcal{K}_{21} & \mathcal{K}_{22} \end{bmatrix}
$$

have the form

$$(\mathcal{K}_l)_{i,j} = \int_{\Omega_2} \nu_2 \nabla\psi_i \cdot \nabla\psi_j \, d\xi, \quad i,j = 1,\ldots,n_a.$$

Here, the basis functions ψ_i are sorted as the state components, i.e.,

$$\begin{aligned} &\operatorname{supp}(\psi_i) \cap \Omega_1 \neq \emptyset, && i = 1,\ldots,n_1, \\ &\operatorname{supp}(\psi_i) \cap \Omega_1 = \emptyset, && i = n_1+1,\ldots,n_1+n_2 = n_a. \end{aligned}$$

Since $\mathcal{K}(a)$ is symmetric and positive definite, the block $\mathcal{K}_{22}$ is symmetric and positive definite too.

4.3.3 Index analysis and index reduction

Here, we investigate the tractability index of the semidiscretized 2D MQS system (4.3.5) and present a transformation of this DAE system into the ODE form. The following theorem establishes that the MQS system has tractability index one. A similar result for the FIT discretized MQS model was obtained in [BBS11].

Theorem 4.5. *Consider a DAE* (4.3.5), (4.3.6), (4.3.12), *where* $\mathcal{M}_{11}$ *and* $\mathcal{K}_{22}$ *being symmetric, positive definite and* $\mathcal{X}_2$ *having full column rank. This system has tractability index one.*

Proof. Let the columns of $\mathcal{Y}$ form an orthonormal basis of $\ker(\mathcal{X}_2^T)$. Then the projector $\mathcal{Q}_0$ onto $\ker(\mathcal{G}_0)$ with $\mathcal{G}_0 = \mathcal{E}$ can be defined as

$$\mathcal{Q}_0 = \begin{bmatrix} 0 & 0 & 0 \\ 0 & \mathcal{Y}\mathcal{Y}^T & 0 \\ 0 & 0 & I_m \end{bmatrix}.$$

For $\mathcal{B}_0(x) = -\frac{\partial}{\partial x}(\mathcal{A}(x)x)$, we have

$$\begin{aligned} \mathcal{G}_1 &= \mathcal{G}_0 + \mathcal{B}_0(x)\mathcal{Q}_0 \\ &= \begin{bmatrix} \mathcal{M}_{11} & \mathcal{K}_{12}\mathcal{Y}\mathcal{Y}^T & -\mathcal{X}_1 \\ 0 & \mathcal{K}_{22}\mathcal{Y}\mathcal{Y}^T & -\mathcal{X}_2 \\ \mathcal{X}_1^T & \mathcal{X}_2^T & \mathcal{R} \end{bmatrix}. \end{aligned}$$

To show that $\mathcal{G}_1$ is invertible, we consider the equation $\mathcal{G}_1[v_1, v_2, v_3]^T = 0$, which is equivalent to

$$\mathcal{M}_{11}v_1 + \mathcal{K}_{12}\mathcal{Y}\mathcal{Y}^T v_2 - \mathcal{X}_1 v_3 = 0, \tag{4.3.14a}$$

$$\mathcal{K}_{22}\mathcal{Y}\mathcal{Y}^T v_2 - \mathcal{X}_2 v_3 = 0, \tag{4.3.14b}$$

$$\mathcal{X}_1^T v_1 + \mathcal{X}_2^T v_2 + \mathcal{R} v_3 = 0. \tag{4.3.14c}$$

Multiplying equation (4.3.14b) from the left with $\mathcal{Y}^T$ and using that $\mathcal{Y}^T\mathcal{X}_2 = 0$, we obtain $\mathcal{Y}^T\mathcal{K}_{22}\mathcal{Y}\mathcal{Y}^T v_2 = 0$. Since $\mathcal{K}_{22}$ is symmetric, positive definite and $\mathcal{Y}$ has full column rank, $\mathcal{Y}^T v_2 = 0$. Equation (4.3.14b), therefore, gives $\mathcal{X}_2 v_3 = 0$. From the full column rank condition for $\mathcal{X}_2$ follows $v_3 = 0$. We now insert $v_3 = 0$ and $\mathcal{Y}^T v_2 = 0$ in equation (4.3.14a). Then the invertibility of $\mathcal{M}_{11}$ gives $v_1 = 0$. The last equation (4.3.14c) reads then $\mathcal{X}_2^T v_2 = 0$. This means that v_2 belongs to the image of $\mathcal{Y}$ and the kernel of $\mathcal{Y}^T$ and, therefore, $v_2 = 0$. Thus, $[v_1^T, v_2^T, v_3^T]^T = 0$ and, hence, $\mathcal{G}_1$ is invertible. This implies that (4.3.5) has tractability index one. □

In the following, we perform an index reduction to the DAE system (4.3.5) by finding an ODE formulation

$$\mathcal{E}_1\dot{x}_1 = \mathcal{A}_1(x_1)x_1 + \mathcal{B}_1 u, \tag{4.3.15a}$$
$$y = \mathcal{C}_1 x_1 \tag{4.3.15b}$$

with a new state x_1 and system matrices $\mathcal{E}_1$, $\mathcal{A}_1(x_1)$, $\mathcal{B}_1$ and $\mathcal{C}_1$, where $\mathcal{E}_1$ is nonsingular. For this purpose, we consider the matrix $\mathcal{Y}$ as in the proof above and define $\mathcal{Z} = \mathcal{X}_2(\mathcal{X}_2^T\mathcal{X}_2)^{-\frac{1}{2}}$ whose columns span $\operatorname{im}(\mathcal{X}_2)$. Then it follows from $\mathcal{Z}^T\mathcal{Z} = I$ and $\mathcal{Z}^T\mathcal{Y} = 0$ that the matrix

$$\mathcal{T} = \begin{bmatrix} I_{n_1} & 0 & 0 \\ 0 & \mathcal{Z}^T & 0 \\ 0 & \mathcal{Y}^T & 0 \\ 0 & 0 & I_m \end{bmatrix} \tag{4.3.16}$$

is nonsingular and $\mathcal{T}^{-1} = \mathcal{T}^T$. We now introduce a new state vector

$$\begin{bmatrix} a_1 \\ a_{21} \\ a_{22} \\ \iota \end{bmatrix} = \mathcal{T}\begin{bmatrix} a \\ \iota \end{bmatrix}, \tag{4.3.17}$$

insert it in equation (4.3.5) and multiply this equation from left with $\mathcal{T}$. Then the transformed DAE can be written as

$$\begin{aligned} \mathcal{M}_{11}\dot{a}_1 &= -\mathcal{K}_{11}(a_1)a_1 - \mathcal{K}_{12}\mathcal{Z}a_{21} - \mathcal{K}_{12}\mathcal{Y}a_{22} + \mathcal{X}_1\iota, \\ 0 &= -\mathcal{Z}^T\mathcal{K}_{21}a_1 - \mathcal{Z}^T\mathcal{K}_{22}\mathcal{Z}a_{21} - \mathcal{Z}^T\mathcal{K}_{22}\mathcal{Y}a_{22} + \mathcal{Z}^T\mathcal{X}_2\iota, \\ 0 &= -\mathcal{Y}^T\mathcal{K}_{21}a_1 - \mathcal{Y}^T\mathcal{K}_{22}\mathcal{Z}a_{21} - \mathcal{Y}^T\mathcal{K}_{22}\mathcal{Y}a_{22}, \\ \mathcal{X}_1^T\dot{a}_1 + \mathcal{X}_2^T\mathcal{Z}\dot{a}_{21} &= -\mathcal{R}\iota + u. \end{aligned} \tag{4.3.18}$$

Since $\mathcal{Y}^T\mathcal{K}_{22}\mathcal{Y}$ and $\mathcal{R}$ are both nonsingular, we can solve the third and forth equations for a_{22} and ι and get

$$a_{22} = -(\mathcal{Y}^T\mathcal{K}_{22}\mathcal{Y})^{-1}\mathcal{Y}^T\mathcal{K}_{21}a_1 - (\mathcal{Y}^T\mathcal{K}_{22}\mathcal{Y})^{-1}\mathcal{Y}^T\mathcal{K}_{22}\mathcal{Z}a_{21}, \tag{4.3.19a}$$
$$\iota = -\mathcal{R}^{-1}\mathcal{X}_1^T\dot{a}_1 - \mathcal{R}^{-1}\mathcal{X}_2^T\mathcal{Z}\dot{a}_{21} + \mathcal{R}^{-1}u. \tag{4.3.19b}$$

Substituting these vectors into the first and second equations in (4.3.18) leads to the state equation (4.3.15) with matrices

$$\mathcal{E}_1 = \begin{bmatrix} \mathcal{M}_{11} + \mathcal{X}_1\mathcal{R}^{-1}\mathcal{X}_1^T & \mathcal{X}_1\mathcal{R}^{-1}\mathcal{X}_2^T\mathcal{Z} \\ \mathcal{Z}^T\mathcal{X}_2\mathcal{R}^{-1}\mathcal{X}_1^T & \mathcal{Z}^T\mathcal{X}_2\mathcal{R}^{-1}\mathcal{X}_2^T\mathcal{Z} \end{bmatrix}, \tag{4.3.20a}$$

$$\begin{aligned} \mathcal{A}_1(x_1) = &-\begin{bmatrix} \mathcal{K}_{11}(a_1) & \mathcal{K}_{12}\mathcal{Z} \\ \mathcal{Z}^T\mathcal{K}_{21} & \mathcal{Z}^T\mathcal{K}_{22}\mathcal{Z} \end{bmatrix} \\ &+ \begin{bmatrix} \mathcal{K}_{12} \\ \mathcal{Z}^T\mathcal{K}_{22} \end{bmatrix}\mathcal{Y}(\mathcal{Y}^T\mathcal{K}_{22}\mathcal{Y})^{-1}\mathcal{Y}^T\begin{bmatrix} \mathcal{K}_{21} & \mathcal{K}_{22}\mathcal{Z} \end{bmatrix}, \end{aligned} \tag{4.3.20b}$$

$$\mathcal{B}_1 = \begin{bmatrix} \mathcal{X}_1 \\ \mathcal{Z}^T\mathcal{X}_2 \end{bmatrix}\mathcal{R}^{-1} \tag{4.3.20c}$$

and the state $x_1 = \left[a_1^T, a_{21}^T\right]^T$.

To calculate the output equation, we need the time derivative of the solution. The derivative can be determined from the state equation provided $\mathcal{E}_1$ is nonsingular. This property is established in the following theorem.

Theorem 4.6. *Let $\mathcal{M}_{11}$ and $\mathcal{R}$ be nonsingular and let $\mathcal{X}_2$ have full column rank. Then the matrix $\mathcal{E}_1$ in (4.3.20a) is nonsingular and its inverse is given by*

$$\mathcal{E}_1^{-1} = \begin{bmatrix} \mathcal{M}_{11}^{-1} & -\mathcal{M}_{11}^{-1}\mathcal{X}_1(\mathcal{X}_2^T\mathcal{X}_2)^{-\frac{1}{2}} \\ -(\mathcal{X}_2^T\mathcal{X}_2)^{-\frac{1}{2}}\mathcal{X}_1^T\mathcal{M}_{11}^{-1} & (\mathcal{X}_2^T\mathcal{X}_2)^{-\frac{1}{2}}(\mathcal{R}+\mathcal{X}_1^T\mathcal{M}_{11}^{-1}\mathcal{X}_1)(\mathcal{X}_2^T\mathcal{X}_2)^{-\frac{1}{2}} \end{bmatrix}.$$

Proof. Using $\mathcal{X}_2^T\mathcal{Z} = (\mathcal{X}_2^T\mathcal{X}_2)^{\frac{1}{2}}$, the matrix $\mathcal{E}_1$ can be factorized as

$$\mathcal{E}_1 = \begin{bmatrix} \mathcal{M}_{11} & \mathcal{X}_1\mathcal{R}^{-1}(\mathcal{X}_2^T\mathcal{X}_2)^{\frac{1}{2}} \\ 0 & (\mathcal{X}_2^T\mathcal{X}_2)^{\frac{1}{2}}\mathcal{R}^{-1}(\mathcal{X}_2^T\mathcal{X}_2)^{\frac{1}{2}} \end{bmatrix} \begin{bmatrix} I & 0 \\ (\mathcal{X}_2^T\mathcal{X}_2)^{-\frac{1}{2}}\mathcal{X}_1^T & I \end{bmatrix}.$$

Since $\mathcal{M}_{11}$ and $(\mathcal{X}_2^T\mathcal{X}_2)^{\frac{1}{2}}\mathcal{R}^{-1}(\mathcal{X}_2^T\mathcal{X}_2)^{\frac{1}{2}}$ are both nonsingular, the matrix $\mathcal{E}_1$ is nonsingular too. Its inverse is given by

$$\begin{aligned}\mathcal{E}_1^{-1} &= \begin{bmatrix} I & 0 \\ -(\mathcal{X}_2^T\mathcal{X}_2)^{-\frac{1}{2}}\mathcal{X}_1^T & I \end{bmatrix} \begin{bmatrix} \mathcal{M}_{11}^{-1} & -\mathcal{M}_{11}^{-1}\mathcal{X}_1(\mathcal{X}_2^T\mathcal{X}_2)^{-\frac{1}{2}} \\ 0 & (\mathcal{X}_2^T\mathcal{X}_2)^{-\frac{1}{2}}\mathcal{R}(\mathcal{X}_2^T\mathcal{X}_2)^{-\frac{1}{2}} \end{bmatrix} \\ &= \begin{bmatrix} \mathcal{M}_{11}^{-1} & -\mathcal{M}_{11}^{-1}\mathcal{X}_1(\mathcal{X}_2^T\mathcal{X}_2)^{-\frac{1}{2}} \\ -(\mathcal{X}_2^T\mathcal{X}_2)^{-\frac{1}{2}}\mathcal{X}_1^T\mathcal{M}_{11}^{-1} & (\mathcal{X}_2^T\mathcal{X}_2)^{-\frac{1}{2}}(\mathcal{R}+\mathcal{X}_1^T\mathcal{M}_{11}^{-1}\mathcal{X}_1)(\mathcal{X}_2^T\mathcal{X}_2)^{-\frac{1}{2}} \end{bmatrix}.\end{aligned}$$

□

We now compute the output function. Using equations (4.3.19b) and (4.3.20c), the state equation (4.3.15a) and the inverse of $\mathcal{E}_1$, we obtain

$$\begin{aligned} y = \iota &= -\mathcal{R}^{-1}\begin{bmatrix}\mathcal{X}_1^T & \mathcal{X}_2^T\mathcal{Z}\end{bmatrix}\begin{bmatrix}\dot{a}_1 \\ \dot{a}_{21}\end{bmatrix} + \mathcal{R}^{-1}u \\ &= -\mathcal{R}^{-1}\begin{bmatrix}\mathcal{X}_1^T & \mathcal{X}_2^T\mathcal{Z}\end{bmatrix}\mathcal{E}_1^{-1}(\mathcal{A}_1(x_1)x_1 + \mathcal{B}_1 u) + \mathcal{R}^{-1}u \\ &= -\mathcal{B}_1^T\mathcal{E}_1^{-1}\mathcal{A}_1(x_1)x_1 - (\mathcal{B}_1^T\mathcal{E}_1^{-1}\mathcal{B}_1 - \mathcal{R}^{-1})u \\ &= \mathcal{C}_1 x_1 \end{aligned}$$

with

$$\begin{aligned} \mathcal{C}_1 &= -\mathcal{B}_1^T\mathcal{E}_1^{-1}\mathcal{A}_1(x_1) \\ &= \begin{bmatrix}0 & -(\mathcal{X}_2^T\mathcal{X}_2)^{-\frac{1}{2}}\end{bmatrix}\mathcal{A}_1(x_1) \\ &= \begin{bmatrix}(\mathcal{X}_2^T\mathcal{X}_2)^{-1}\mathcal{X}_2^T\mathcal{K}_{21} & (\mathcal{X}_2^T\mathcal{X}_2)^{-1}\mathcal{X}_2^T\mathcal{K}_{22}\mathcal{Z}\end{bmatrix} \\ &\quad - (\mathcal{X}_2^T\mathcal{X}_2)^{-1}\mathcal{X}_2^T\mathcal{K}_{22}\mathcal{Y}(\mathcal{Y}^T\mathcal{K}_{22}\mathcal{Y})^{-1}\mathcal{Y}^T\begin{bmatrix}\mathcal{K}_{21} & \mathcal{K}_{22}\mathcal{Z}\end{bmatrix} \\ &= (\mathcal{X}_2^T\mathcal{X}_2)^{-1}\mathcal{X}_2^T(I - \mathcal{K}_{22}\mathcal{Y}(\mathcal{Y}^T\mathcal{K}_{22}\mathcal{Y})^{-1}\mathcal{Y}^T)\begin{bmatrix}\mathcal{K}_{21} & \mathcal{K}_{22}\mathcal{Z}^T\end{bmatrix}. \end{aligned} \tag{4.3.21}$$

Here, we used the equation

$$\mathcal{B}_1^T\mathcal{E}_1^{-1}\mathcal{B}_1 - \mathcal{R}^{-1} = \begin{bmatrix}0 & (\mathcal{X}_2^T\mathcal{X}_2)^{-\frac{1}{2}}\end{bmatrix}\mathcal{B}_1 - \mathcal{R}^{-1} = \mathcal{R}^{-1} - \mathcal{R}^{-1} = 0.$$

Remark 4.7. In [BCS12], the MQS system (4.3.5) was transformed into an ODE

$$\begin{aligned} \breve{\mathcal{E}}_1 \dot{\breve{x}} &= \breve{\mathcal{A}}_1(\breve{x})\breve{x} + \breve{\mathcal{B}}_1 u, \\ y &= \breve{\mathcal{C}}_1 \breve{x} \end{aligned} \tag{4.3.22}$$

with $\breve{x} = \left[a_1^T, \iota^T\right]^T$ and

$$\begin{aligned} \breve{\mathcal{E}}_1 &= \begin{bmatrix} \mathcal{M}_{11} & 0 \\ X_1^T - X_2^T \mathcal{K}_{22}^{-1} \mathcal{K}_{21}^T & X_2^T \mathcal{K}_{22}^{-1} X_2 \end{bmatrix}, \\ \breve{\mathcal{A}}_1 &= \begin{bmatrix} -\mathcal{K}_{11}(a_1) + \mathcal{K}_{12}\mathcal{K}_{22}^{-1}\mathcal{K}_{21} & X_1^T - \mathcal{K}_{12}\mathcal{K}_{22}^{-1} X_2 \\ 0 & -\mathcal{R} \end{bmatrix}, \\ \breve{\mathcal{B}}_1 &= \begin{bmatrix} 0 \\ I \end{bmatrix} = \breve{\mathcal{C}}_1^T \end{aligned}$$

which, unlike the ODE (4.3.15), involves the derivative of the current vector ι. System (4.3.22) together with the equation

$$a_2 = -\mathcal{K}_{22}^{-1}\mathcal{K}_{21} a_1 + \mathcal{K}_{22}^{-1} X_2 \iota \tag{4.3.23}$$

is equivalent to the DAE (4.3.5). Note that differentiating (4.3.23) and using the state equation in (4.3.22), we obtain the underlying ODE for system (4.3.5). This shows that (4.3.22) has also differentiation index one.

4.3.4 Passivity

In this section, we investigate passivity of the variational MQS system (4.3.3), (4.3.2b), (4.3.2d) with the output $y = \iota$, the semidiscretized system (4.3.5), (4.3.6) and the ODE system (4.3.15), (4.3.20), (4.3.21). Passivity for the variational problem (4.3.3) can be defined analogously to the finite-dimensional case, see Definition 2.13.

Definition 4.8. The variational MQS problem (4.3.3), (4.3.2b), (4.3.2d) with the output $y = \iota$ is called *passive* if there exists a nonnegative function $S : H_0^1(\Omega) \to \mathbb{R}_0^+$ such that $S(0) = 0$ and for all $T > 0$ and all quadratically integrable inputs v admissible with the initial condition $\phi(\cdot, 0) = \phi_0(\cdot)$ the passivation inequality

$$S(\phi(\cdot, T)) - S(\phi_0(\cdot)) \leqslant \int_0^T y^T(\tau) v(\tau)\, d\tau \tag{4.3.24}$$

is satisfied, where ϕ is the weak solution of (4.3.3), (4.3.2b), (4.3.2d). The function S is called *storage function*.

The following theorem shows that the variational MQS problem is passive.

Theorem 4.9. *The variational MQS system* (4.3.3), (4.3.2b), (4.3.2d) *with the output* $y = \iota$ *is passive.*

Proof. Consider a function $\vartheta : \Omega \times \mathbb{R}_0^+ \to \mathbb{R}_0^+$ given by

$$\vartheta(\xi, \varrho) = \frac{1}{2} \int_0^{\varrho} \boldsymbol{\nu}(\xi, \sqrt{\sigma})\, d\sigma = \int_0^{\sqrt{\varrho}} \boldsymbol{\nu}(\xi, \sigma)\sigma\, d\sigma \tag{4.3.25}$$

and define a storage function as

$$S(\phi(\cdot, t)) = \int_\Omega \vartheta(\xi, \|\nabla\phi(\xi, t)\|^2)\, d\xi.$$

This function is nonnegative, since $\boldsymbol{\nu}$ is positive. Furthermore, we have $S(0) = 0$. The continuous differentiability follows from the differentiability of ϑ in the second argument and the differentiability of the norm. We now show that

$$\frac{d}{dt} S(\phi(\cdot, t)) \leqslant y(t)^T v(t)$$

for all v and suitable ϕ, $y = \iota$ that satisfy (4.3.3). We calculate

$$\begin{aligned} \frac{d}{dt} S(\phi(\cdot, t)) &= \frac{d}{dt} \int_\Omega \vartheta(\xi, \|\nabla\phi(\xi, t)\|^2)\, d\xi \\ &= \int_\Omega \frac{\partial}{\partial\varrho} \vartheta(\xi, \|\nabla\phi(\xi, t)\|^2) \frac{\partial}{\partial t} \|\nabla\phi(\xi, t)\|^2\, d\xi \\ &= \int_\Omega \boldsymbol{\nu}(\xi, \|\nabla\phi(\xi, t)\|) \nabla\phi(\xi, t) \cdot \frac{\partial}{\partial t} \nabla\phi(\xi, t)\, d\xi. \end{aligned}$$

Taking ϕ as a test function and $\frac{\partial}{\partial t}\phi$ as a trial function, we obtain using equations (4.3.3) and (4.3.2b) that

$$\begin{aligned} \frac{d}{dt} S(\phi(\cdot, t)) &= -\int_\Omega \boldsymbol{\sigma} \frac{\partial}{\partial t}\phi(\xi, t) \frac{\partial}{\partial t}\phi(\xi, t)\, d\xi + \int_\Omega \frac{\partial}{\partial t}\phi(\xi, t) \boldsymbol{\chi}_{\mathrm{str},3}\, d\xi\, \iota \\ &= -\int_\Omega \boldsymbol{\sigma} \frac{\partial}{\partial t}\phi(\xi, t) \frac{\partial}{\partial t}\phi(\xi, t)\, d\xi - \iota^T(t) \mathcal{R} \iota(t) + v^T(t)\iota(t) \\ &\leqslant y^T(t) v(t). \end{aligned}$$

Here, we used the property that the first two summands are negative, since $\boldsymbol{\sigma}$ is positive on Ω and $\mathcal{R}$ is positive definite. Integrating this inequality on $[0, T]$, we get the passivation inequality (4.3.24) and, hence, (4.3.3), (4.3.2b) is passive. □

The following theorem establishes that the spatial discretization of the variational MQS problem (4.3.3), (4.3.2b), (4.3.2d) as described in the previous section preserves passivity.

Theorem 4.10. *The semidiscretized MQS system* (4.3.5), (4.3.6) *is passive.*

Proof. Similar to the variational problem, we define a storage function

$$S_d(\boldsymbol{a}(t)) := \int_\Omega \vartheta(\xi, \|\nabla \sum_{i=1}^{n_a} \alpha_i(t)\psi_i(\xi)\|^2)\, d\xi,$$

where ϑ is given in (4.3.25) and $\boldsymbol{a} = \left[\alpha_1, \dots, \alpha_{n_a}\right]^T$. This function is nonnegative, since $\boldsymbol{\nu}$ is positive and $S_d(0) = 0$ due to the definition of ϑ. The calculation of the time derivative of S_d results in

$$\begin{aligned}
&\frac{d}{dt} S_d(\boldsymbol{a}(t)) = \frac{d}{dt} \int_\Omega \vartheta(\xi, \|\nabla \sum_{i=1}^{n_a} \alpha_i(t)\psi_i(\xi)\|^2)\, d\xi \\
&= \int_\Omega \frac{\partial}{\partial \varrho} \vartheta(\xi, \|\nabla \sum_{i=1}^{n_a} \alpha_i(t)\psi_i(\xi)\|^2) \frac{\partial}{\partial t} \|\nabla \sum_{i=1}^{n_a} \alpha_i(t)\psi_i(\xi)\|^2\, d\xi \\
&= \int_\Omega \boldsymbol{\nu}(\xi, \|\nabla \sum_{i=1}^{n_a} \alpha_i(t)\psi_i(\xi)\|) \left(\nabla \sum_{j=1}^{n_a} \alpha_j(t)\psi_j(\xi)\right) \cdot \left(\nabla \sum_{k=1}^{n_a} \dot{\alpha}_k(t)\psi_k(\xi)\right) d\xi \\
&= \sum_{j=1}^{n_a} \alpha_j(t) \sum_{k=1}^{n_a} \dot{\alpha}_k(t) \int_\Omega \boldsymbol{\nu}(\xi, \|\nabla \sum_{i=1}^{n_a} \alpha_i(t)\psi_i(\xi)\|) \nabla\psi_j(\xi) \cdot \nabla\psi_k(\xi)\, d\xi \\
&= \dot{\boldsymbol{a}}^T(t) \mathcal{K}(\boldsymbol{a}(t)) \boldsymbol{a}(t).
\end{aligned}$$

It follows from (4.3.5) that

$$\begin{aligned}
\frac{d}{dt} S_d(\boldsymbol{a}(t)) &= -\dot{\boldsymbol{a}}^T(t) \mathcal{M} \dot{\boldsymbol{a}}(t) + \dot{\boldsymbol{a}}^T(t) \mathcal{X} \iota(t) \\
&= -\dot{\boldsymbol{a}}^T(t) \mathcal{M} \dot{\boldsymbol{a}}(t) - \iota^T(t) \mathcal{R} \iota(t) + u^T(t)\iota(t) \\
&\leqslant y^T(t) u(t).
\end{aligned}$$

Then integrating this inequality in $[0, T]$, we obtain

$$S_d(\boldsymbol{a}(T)) - S_d(\boldsymbol{a}(0)) \leqslant \int_0^T y^T(t) u(t)\, dt.$$

Thus, system (4.3.5) is passive. □

Finally, we show that the ODE system (4.3.15) inherit passivity of the DAE system (4.3.5).

Theorem 4.11. *The ODE system* (4.3.15), (4.3.20), (4.3.21) *is passive.*

Proof. The result can be proved analogously to Theorem 4.10. Using (4.3.16), (4.3.17) and the third equation in (4.3.18), we find

$$\begin{aligned}
\boldsymbol{a} &= \begin{bmatrix} \boldsymbol{a}_1 \\ \boldsymbol{a}_2 \end{bmatrix} = \begin{bmatrix} \boldsymbol{a}_1 \\ \mathcal{Z}\boldsymbol{a}_{21} + \mathcal{Y}\boldsymbol{a}_{22} \end{bmatrix} \\
&= \begin{bmatrix} \boldsymbol{a}_1 \\ \mathcal{Z}\boldsymbol{a}_{21} - \mathcal{Y}(\mathcal{Y}^T \mathcal{K}_{22} \mathcal{Y})^{-1} \mathcal{Y}^T (\mathcal{K}_{21}\boldsymbol{a}_1 + \mathcal{K}_{22}\mathcal{Z}\boldsymbol{a}_{21}) \end{bmatrix} = R \begin{bmatrix} \boldsymbol{a}_1 \\ \boldsymbol{a}_{21} \end{bmatrix},
\end{aligned} \tag{4.3.26}$$

where

$$R = \begin{bmatrix} I_{n_1} & 0 \\ -\mathcal{Y}(\mathcal{Y}^T \mathcal{K}_{22}\mathcal{Y})^{-1}\mathcal{Y}^T \mathcal{K}_{21} & (I_{n_2} - \mathcal{Y}(\mathcal{Y}^T \mathcal{K}_{22}\mathcal{Y})^{-1}\mathcal{Y}^T \mathcal{K}_{22})\mathcal{Z} \end{bmatrix}.$$

Introducing new basis functions

$$[\phi_1(\xi), \ldots, \phi_{n_1+m}(\xi)] = [\psi_1(\xi), \ldots, \psi_{n_a}(\xi)]R,$$

we obtain

$$\sum_{i=1}^{n_a} \alpha_i(t)\psi_i(\xi) = \sum_{i=1}^{n_1+m} \beta_i(t)\phi_i(\xi),$$

where

$$[\beta_1(t), \ldots, \beta_{n_1}(t)]^T = \boldsymbol{a}_1(t) = [\alpha_1(t), \ldots, \alpha_{n_1}(t)]^T$$

and

$$[\beta_{n_1+1}(t), \ldots, \beta_{n_1+m}(t)]^T = \boldsymbol{a}_{21}(t).$$

For $x_1(t) = [\beta_1(t), \ldots, \beta_{n_1+m}(t)]^T$, we define a storage function

$$S_1(x_1(t)) = \int_\Omega \vartheta(\xi, \|\nabla \sum_{i=1}^{n_1+m} \beta_i(t)\phi_i(\xi)\|^2)\, d\xi,$$

with ϑ as in (4.3.25). This function is nonnegative, since $\boldsymbol{\nu}$ is positive and $S_1(0) = 0$ due to the definition of ϑ. We calculate

$$\begin{aligned}
\frac{d}{dt} S_1(x_1(t)) &= \frac{d}{dt} \int_\Omega \vartheta(\xi, \|\nabla \sum_{i=1}^{n_1+m} \beta_i(t)\phi_i(\xi)\|^2)\, d\xi \\
&= \int_\Omega \frac{\partial}{\partial \varrho}\vartheta(\xi, \|\nabla \sum_{i=1}^{n_1+m} \beta_i(t)\phi_i(\xi)\|^2) \frac{\partial}{\partial t} \|\nabla \sum_{i=1}^{n_1+m} \beta_i(t)\phi_i(\xi)\|^2\, d\xi \\
&= \int_\Omega \boldsymbol{\nu}(\xi, \|\nabla \sum_{i=1}^{n_1+m} \beta_i(t)\phi_i(\xi)\|) \left(\nabla \sum_{i=1}^{n_1+m} \beta_i(t)\phi_i(\xi)\right) \cdot \left(\nabla \sum_{i=1}^{n_1+m} \dot{\beta}_i(t)\phi_i(\xi)\right) d\xi \\
&= \int_\Omega \boldsymbol{\nu}(\xi, \|\nabla \sum_{i=1}^{n_a} \alpha_i(t)\psi_i(\xi)\|) \left(\nabla \sum_{i=1}^{n_a} \alpha_i(t)\psi_i(\xi)\right) \cdot \left(\nabla \sum_{i=1}^{n_a} \dot{\alpha}_i(t)\psi_i(\xi)\right) d\xi \\
&= \dot{\boldsymbol{a}}^T(t)\mathcal{K}(\boldsymbol{a}(t))\boldsymbol{a}(t).
\end{aligned}$$

Using the relations

$$\begin{aligned}
\mathcal{A}_1(x_1) &= -R^T \mathcal{K}(Rx_1)R, \\
\mathcal{E}_1 &= R^T(\mathcal{M} + X\mathcal{R}^{-1}X)R \\
&= \begin{bmatrix} \mathcal{M}_{11} & 0 \\ 0 & 0 \end{bmatrix} + \mathcal{B}_1\mathcal{R}\mathcal{B}_1^T,
\end{aligned}$$

and equation (4.3.15), we can continue

$$\begin{aligned}\frac{d}{dt}S_1(x_1(t)) &= -\dot{x}_1^T(t)\mathcal{A}_1(x_1(t))x_1(t)\\ &= -\dot{x}_1^T(t)\mathcal{E}_1\dot{x}_1(t) + \dot{x}_1^T(t)\mathcal{B}_1u(t)\\ &= -\dot{a}_1^T(t)\mathcal{M}_{11}\dot{a}_1(t) - \dot{x}_1^T(t)\mathcal{B}_1\mathcal{R}\mathcal{B}_1^T\dot{x}_1(t) + \dot{x}_1^T(t)\mathcal{B}_1u(t)\\ &= -\dot{a}_1^T(t)\mathcal{M}_{11}\dot{a}_1(t) + \dot{x}_1^T(t)\mathcal{B}_1(u(t) - \mathcal{R}\mathcal{B}_1^T\dot{x}_1(t)).\end{aligned}$$

Since the matrix $\mathcal{M}_{11}$ is positive definite, the first summand is negative. Furthermore, we use the output equation $y = -\mathcal{B}_1^T\dot{x}_1 + \mathcal{R}^{-1}u$ twice and obtain

$$\begin{aligned}\frac{d}{dt}S_1(x_1(t)) &\leqslant (\mathcal{R}^{-1}u(t) - y(t))^T\mathcal{R}(\mathcal{R}^{-1}u(t) + y(t) - \mathcal{R}^{-1}u(t))\\ &= -y^T(t)\mathcal{R}y(t) + y^T(t)u(t)\\ &\leqslant y^T(t)u(t).\end{aligned}$$

Integrating this inequality on $[0, T]$, we get the passivation inequality

$$S_1(x_1(T)) - S_1(x_1(0)) \leqslant \int_0^T y^T(t)u(t)\,dt$$

which implies the passivity of the ODE system (4.3.15). □

It follows from Remark 2.17 and Theorems 4.10 and 4.11 that the MQS system (4.3.5) and its ODE form (4.3.15) are io-passive. For linear systems, this property can equivalently be characterized by positive realness of the transfer function. Assume that $\boldsymbol{\nu}$ is constant on Ω_1. Then the matrix $\mathcal{K}$ in (4.3.6) does not depend on the semidiscretized potential $\boldsymbol{a}$, and system (4.3.5) can be transformed to the ODE system (4.3.15) with constant matrices. First, we show that the matrices $\mathcal{E}_1$ and $-\mathcal{A}_1$ are positive definite.

Theorem 4.12. *If $\mathcal{M}_{11}$, $\mathcal{K}$ and $\mathcal{R}$ are symmetric and positive definite, then $\mathcal{E}_1$ and $-\mathcal{A}_1$ in* (4.3.20a) *and* (4.3.20b), *respectively, are symmetric and positive definite.*

Proof. The matrix $\mathcal{E}_1$ can be decomposed as

$$\mathcal{E}_1 = \mathcal{T}_1^T\begin{bmatrix}\mathcal{M}_{11} & 0\\ 0 & (X_2^TX_2)^{\frac{1}{2}}\mathcal{R}^{-1}(X_2^TX_2)^{\frac{1}{2}}\end{bmatrix}\mathcal{T}_1$$

with the nonsingular matrix

$$\mathcal{T}_1 = \begin{bmatrix}I & 0\\ (X_2^TX_2)^{-\frac{1}{2}}X_1^T & I\end{bmatrix}.$$

Since $\mathcal{M}_{11}$ and $\mathcal{R}$ are both symmetric and positive definite, $\mathcal{E}_1$ is symmetric and positive definite too.

The matrix $\mathcal{A}_1$ is the Schur complement of the symmetric, negative definite matrix

$$-\mathcal{T}_2^T\begin{bmatrix}\mathcal{K}_{11} & \mathcal{K}_{12}\\ \mathcal{K}_{21} & \mathcal{K}_{22}\end{bmatrix}\mathcal{T}_2 = -\left[\begin{array}{cc|c}\mathcal{K}_{11} & \mathcal{K}_{12}\mathcal{Z} & \mathcal{K}_{12}\mathcal{Y}\\ \mathcal{Z}^T\mathcal{K}_{21} & \mathcal{Z}^T\mathcal{K}_{22}\mathcal{Z} & \mathcal{Z}^T\mathcal{K}_{22}\mathcal{Y}\\ \hline \mathcal{Y}^T\mathcal{K}_{21} & \mathcal{Y}^T\mathcal{K}_{22}\mathcal{Z} & \mathcal{Y}^T\mathcal{K}_{22}\mathcal{Y}\end{array}\right] \tag{4.3.27}$$

with the orthogonal matrix

$$\mathcal{T}_2 = \begin{bmatrix} I & 0 & 0 \\ 0 & \mathcal{Z} & \mathcal{Y} \end{bmatrix}. \tag{4.3.28}$$

Therefore, $-\mathcal{A}_1$ is symmetric and positive definite, see [HZ05]. □

Unfortunately, the ODE system (4.3.15) does not satisfy Theorem 2.15 providing sufficient conditions for io-passivity. Nevertheless, we can show that its transfer function is positive real implying due to Theorem 2.15 that system (4.3.15) is io-passive.

Theorem 4.13. *If $\mathcal{M}_{11}$, $\mathcal{K}$ and $\mathcal{R}$ are symmetric and positive definite, then system* (4.3.15) *with constant matrices is positive real.*

Proof. It follows from Theorem 4.12 that $\mathcal{E}_1$ and $-\mathcal{A}_1$ are symmetric, positive definite. Then the pencil $\lambda\mathcal{E}_1 - \mathcal{A}_1$ is stable, and, hence, $\mathbf{G}(s) = \mathcal{C}_1(s\mathcal{E}_1 - \mathcal{A}_1)^{-1}\mathcal{B}_1$ is analytic in $\mathbb{C}_+$. Using $\mathcal{C}_1 = -\mathcal{B}_1^T\mathcal{E}_1^{-1}\mathcal{A}_1$, we can compute

$$\begin{aligned}
\mathbf{G}(s) + \mathbf{G}^*(s) &= \mathcal{C}_1(s\mathcal{E}_1 - \mathcal{A}_1)^{-1}\mathcal{B}_1 + \mathcal{B}_1^T(\bar{s}\mathcal{E}_1^T - \mathcal{A}_1^T)^{-1}\mathcal{C}_1^T \\
&= -\mathcal{B}_1^T\mathcal{E}_1^{-1}\mathcal{A}_1(s\mathcal{E}_1 - \mathcal{A}_1)^{-1}\mathcal{B}_1 - \mathcal{B}_1^T(\bar{s}\mathcal{E}_1 - \mathcal{A}_1)^{-1}\mathcal{A}_1\mathcal{E}_1^{-1}\mathcal{B}_1 \\
&= F^*(s)(-\mathcal{E}_1\mathcal{A}_1^{-1}(\bar{s}\mathcal{E}_1 - \mathcal{A}_1) - (s\mathcal{E}_1 - \mathcal{A}_1)\mathcal{A}_1^{-1}\mathcal{E}_1)F(s) \\
&= 2F^*(s)(\mathrm{Re}(s)\mathcal{E}_1(-\mathcal{A}_1)^{-1}\mathcal{E}_1 + \mathcal{E}_1)F(s)
\end{aligned}$$

with $F(s) = \mathcal{E}_1^{-1}\mathcal{A}_1(s\mathcal{E}_1 - \mathcal{A}_1)^{-1}\mathcal{B}_1$. Since the matrix $\mathrm{Re}(s)\mathcal{E}_1(-\mathcal{A}_1)^{-1}\mathcal{E}_1 + \mathcal{E}_1$ is positive definite for all $s \in \mathbb{C}_+$, we have $\mathbf{G}(s) + \mathbf{G}(s)^* \geqslant 0$ for all $s \in \mathbb{C}_+$. Therefore, the transfer function $\mathbf{G}$ is positive real. □

4.4 Model reduction for 2D linear MQS systems

Our model order reduction approach for the linear MQS system is based on the transformation into the ODE form and applying BT to the transformed system. We will show how the special structure of the system matrices can be exploited for the construction of efficient model reduction algorithms avoiding the explicit computation of the ODE system.

Consider the ODE system (4.3.15) with the constant matrices

$$\begin{aligned}
\mathcal{E}_1 &= \begin{bmatrix} \mathcal{M}_{11} + X_1\mathcal{R}^{-1}X_1^T & X_1\mathcal{R}^{-1}X_2^T\mathcal{Z} \\ \mathcal{Z}^TX_2\mathcal{R}^{-1}X_1^T & \mathcal{Z}^TX_2\mathcal{R}^{-1}X_2^T\mathcal{Z} \end{bmatrix}, \\
\mathcal{A}_1 &= -\begin{bmatrix} \mathcal{K}_{11} & \mathcal{K}_{12}\mathcal{Z} \\ \mathcal{Z}^T\mathcal{K}_{21} & \mathcal{Z}^T\mathcal{K}_{22}\mathcal{Z} \end{bmatrix} \\
&\quad + \begin{bmatrix} \mathcal{K}_{12} \\ \mathcal{Z}^T\mathcal{K}_{22} \end{bmatrix} \mathcal{Y}(\mathcal{Y}^T\mathcal{K}_{22}\mathcal{Y})^{-1}\mathcal{Y}^T \begin{bmatrix} \mathcal{K}_{21} & \mathcal{K}_{22}\mathcal{Z} \end{bmatrix}, \\
\mathcal{B}_1 &= \begin{bmatrix} X_1 \\ \mathcal{Z}^TX_2 \end{bmatrix}\mathcal{R}^{-1}, \\
\mathcal{C}_1 &= (X_2^TX_2)^{-1}X_2^T(I - \mathcal{K}_{22}\mathcal{Y}(\mathcal{Y}^T\mathcal{K}_{22}\mathcal{Y})^{-1}\mathcal{Y}^T)\begin{bmatrix} \mathcal{K}_{21} & \mathcal{K}_{22}\mathcal{Z}^T \end{bmatrix}.
\end{aligned} \tag{4.4.1}$$

Our goal is to compute a reduced order model

$$\begin{aligned}
\tilde{\mathcal{E}}_1\dot{\tilde{x}}_1 &= \tilde{\mathcal{A}}_1\tilde{x}_1 + \tilde{\mathcal{B}}_1u, \\
\tilde{y} &= \tilde{\mathcal{C}}_1\tilde{x}_1,
\end{aligned} \tag{4.4.2}$$

where $\tilde{\mathcal{E}}_1, \tilde{\mathcal{A}}_1 \in \mathbb{R}^{\eta\times\eta}$, $\tilde{\mathcal{B}}_1 \in \mathbb{R}^{\eta\times m}$ and $\tilde{\mathcal{C}}_1 \in \mathbb{R}^{m\times\eta}$ with $\eta \ll n_1 + m$. Before we go into detail, we first show a relationship between the Gramians G_c and G_o of system (4.3.15), (4.4.1) which satisfy the generalized Lyapunov equations

$$\mathcal{A}_1 G_c \mathcal{E}_1^T + \mathcal{E}_1 G_c \mathcal{A}_1^T = -\mathcal{B}_1 \mathcal{B}_1^T, \tag{4.4.3}$$
$$\mathcal{A}_1^T G_o \mathcal{E}_1 + \mathcal{E}_1^T G_o \mathcal{A}_1 = -\mathcal{C}_1^T \mathcal{C}_1. \tag{4.4.4}$$

The idea behind this is that the Lyapunov equations are equal for symmetric systems and the ODE system (4.3.15) can be transformed into a symmetric one.

Theorem 4.14. *Let G_c and G_o be the solutions of the Lyapunov equations* (4.4.3) *and* (4.4.4), *respectively. If $\mathcal{M}_{11}$, $\mathcal{K}$ and $\mathcal{R}$ are symmetric and positiv definite, then*

$$\mathcal{E}_1 G_o \mathcal{E}_1 = \mathcal{A}_1 G_c \mathcal{A}_1. \tag{4.4.5}$$

Proof. The left and right multiplication of the Lyapunov equations (4.4.3) and (4.4.4) by $\mathcal{E}_1^{-1}$ and $\mathcal{A}_1^{-1}$, respectively, leads to

$$\mathcal{E}_1^{-1}\mathcal{A}_1 G_c + G_c \mathcal{A}_1 \mathcal{E}_1^{-1} = -\mathcal{E}_1^{-1}\mathcal{B}_1\mathcal{B}_1^T\mathcal{E}_1^{-1},$$
$$G_o \mathcal{E}_1 \mathcal{A}_1^{-1} + \mathcal{A}_1^{-1}\mathcal{E}_1 G_o = -\mathcal{A}_1^{-1}\mathcal{C}_1^T\mathcal{C}_1\mathcal{A}_1^{-1}.$$

Using $\mathcal{C}_1 = -\mathcal{B}_1^T\mathcal{E}_1^{-1}\mathcal{A}_1$ and introducing $G_{cs} = \mathcal{A}_1 G_c \mathcal{A}_1$ and $G_{os} = \mathcal{E}_1 G_o \mathcal{E}_1$, these equations can be written as

$$\mathcal{E}_1^{-1} G_{cs} \mathcal{A}_1^{-1} + \mathcal{A}_1^{-1} G_{cs} \mathcal{E}_1^{-1} = -\mathcal{E}_1^{-1}\mathcal{B}_1\mathcal{B}_1^T\mathcal{E}_1^{-1}, \tag{4.4.6}$$
$$\mathcal{E}_1^{-1} G_{os} \mathcal{A}_1^{-1} + \mathcal{A}_1^{-1} G_{os} \mathcal{E}_1^{-1} = -\mathcal{E}_1^{-1}\mathcal{B}_1\mathcal{B}_1^T\mathcal{E}_1^{-1}. \tag{4.4.7}$$

The pencil $\lambda\mathcal{E}_1^{-1} - \mathcal{A}_1^{-1}$ is stable, because $\mathcal{E}_1^{-1}$ and $-\mathcal{A}_1^{-1}$ are positive definite, and hence, the Lyapunov equations (4.4.6) and (4.4.7) are uniquely solvable. This implies that $G_{cs} = G_{os}$, i.e., (4.4.5) holds. □

Similarly to the standard state space case, see Algorithm 2.1, in the first step of the balanced truncation method, we need to compute the Cholesky factors of the Gramians G_c and G_o. Theorem 4.14 implies that the Cholesky factor Z_o of the observability Gramian $G_o = Z_o Z_o^T$ can be computed from the Cholesky factor Z_c of the controllability Gramian $G_c = Z_c Z_c^T$ as $Z_o = -\mathcal{E}_1^{-1}\mathcal{A}_1 Z_c$. This relation follows from

$$G_o = \mathcal{E}_1^{-1}\mathcal{A}_1 Z_c Z_c^T \mathcal{A}_1 \mathcal{E}_1^{-1} = Z_o Z_o^T.$$

For solving the generalized Lyapunov equation (4.4.3), we use the LR-ADI method given in (2.4.4). In this method, we need to compute $(\tau\mathcal{E}_1 + \mathcal{A}_1)^{-1}v$ for different vectors v depending on the iteration. The matrices $\mathcal{E}_1$ and $\mathcal{A}_1$ are, in general, dense. Therefore, we do not want to calculate $\mathcal{E}_1$ and $\mathcal{A}_1$ explicitly and use the following lemma instead.

Lemma 4.15. *Let $\mathcal{E}_1$ and $\mathcal{A}_1$ be as in* (4.4.1), $\mathcal{Z} = \mathcal{X}_2(\mathcal{X}_2^T\mathcal{X}_2)^{-\frac{1}{2}}$, $\tau \in \mathbb{C}_-$, *and* $[v_1^T, v_2^T]^T \in \mathbb{R}^{n_1+m}$. *Then the vector*

$$z = (\tau\mathcal{E}_1 + \mathcal{A}_1)^{-1}v \tag{4.4.8}$$

Algorithm 4.1: Computation of $(\tau\mathcal{E}_1 + \mathcal{A}_1)^{-1}v$

Input : $\mathcal{M}_{11}, \mathcal{K}_{11} \in \mathbb{R}^{n_1\times n_1}$, $\mathcal{K}_{12} \in \mathbb{R}^{n_1\times n_2}$, $\mathcal{K}_{21} \in \mathbb{R}^{n_2\times n_1}$, $\mathcal{K}_{22} \in \mathbb{R}^{n_2\times n_2}$,
$\mathcal{X}_1 \in \mathbb{R}^{n_1\times m}$, $\mathcal{X}_2 \in \mathbb{R}^{n_2\times m}$, $\mathcal{R} \in \mathbb{R}^{m\times m}$, $v = [v_1^T, v_2^T]^T \in \mathbb{R}^{n_1+m}$, $\tau \in \mathbb{C}_-$.

Output: $z = (\tau\mathcal{E}_1 + \mathcal{A}_1)^{-1}v$ with $\mathcal{E}_1$ and $\mathcal{A}_1$ as given in (4.4.1)

1 Solve the linear system

$$\begin{bmatrix} \tau\mathcal{M}_{11} - \mathcal{K}_{11} & -\mathcal{K}_{12} & \mathcal{X}_1 \\ -\mathcal{K}_{21} & -\mathcal{K}_{22} & \mathcal{X}_2 \\ \tau\mathcal{X}_1^T & \tau\mathcal{X}_2^T & -\mathcal{R} \end{bmatrix} \begin{bmatrix} z_1 \\ z_2 \\ z_3 \end{bmatrix} = \begin{bmatrix} v_1 \\ \mathcal{X}_2(\mathcal{X}_2^T\mathcal{X}_2)^{-\frac{1}{2}}v_2 \\ 0 \end{bmatrix}.$$

2 Compute

$$z = \begin{bmatrix} z_1 \\ (\mathcal{X}_2^T\mathcal{X}_2)^{-\frac{1}{2}}\mathcal{X}_2^T z_2 \end{bmatrix}.$$

---

can be determined as $z = [z_1^T, (\mathcal{Z}^T z_2)^T]^T$, *where* z_1 *and* z_2 *satisfy the sparse linear system*

$$(\tau\mathcal{E} + \mathcal{A}) \begin{bmatrix} z_1 \\ z_2 \\ z_3 \end{bmatrix} = \begin{bmatrix} v_1 \\ \mathcal{Z}v_2 \\ 0 \end{bmatrix} \tag{4.4.9}$$

with $\mathcal{E}$ *and* $\mathcal{A}$ *given in* (4.3.6).

Proof. Using $\mathcal{T}$ as in (4.3.16), we transform equation (4.4.9) to

$$\begin{aligned} (\tau\mathcal{M}_{11} - \mathcal{K}_{11})z_1 - \mathcal{K}_{12}\mathcal{Z}(\mathcal{Z}^T z_2) - \mathcal{K}_{12}\mathcal{Y}(\mathcal{Y}^T z_2) + \mathcal{X}_1 z_3 &= v_1, \\ -\mathcal{Z}^T\mathcal{K}_{21}z_1 - \mathcal{Z}^T\mathcal{K}_{22}\mathcal{Z}(\mathcal{Z}^T z_2) - \mathcal{Z}^T\mathcal{K}_{22}\mathcal{Y}(\mathcal{Y}^T z_2) + \mathcal{Z}^T\mathcal{X}_2 z_3 &= \mathcal{Z}^T\mathcal{Z}v_2, \\ -\mathcal{Y}^T\mathcal{K}_{21}z_1 - \mathcal{Y}^T\mathcal{K}_{22}\mathcal{Z}(\mathcal{Z}^T z_2) - \mathcal{Y}^T\mathcal{K}_{22}\mathcal{Y}(\mathcal{Y}^T z_2) &= 0, \\ \tau\mathcal{X}_1^T z_1 + \tau\mathcal{X}_2^T\mathcal{Z}(\mathcal{Z}^T z_2) - \mathcal{R}z_3 &= 0. \end{aligned}$$

Solving the last two equations for $\mathcal{Y}^T z_2$ and z_3, we get

$$\begin{aligned} \mathcal{Y}^T z_2 &= -(\mathcal{Y}^T\mathcal{K}_{22}\mathcal{Y})^{-1}\begin{bmatrix} \mathcal{Y}^T\mathcal{K}_{21} & \mathcal{Y}^T\mathcal{K}_{22}\mathcal{Z} \end{bmatrix}\begin{bmatrix} z_1 \\ \mathcal{Z}^T z_2 \end{bmatrix}, \\ z_3 &= \tau\mathcal{R}^{-1}\begin{bmatrix} \mathcal{X}_1^T & \mathcal{X}_2^T\mathcal{Z} \end{bmatrix}\begin{bmatrix} z_1 \\ \mathcal{Z}^T z_2 \end{bmatrix}. \end{aligned}$$

Substituting these vectors into the first two equations gives finally equation (4.4.8). □

We summarize the computation of $(\tau\mathcal{E}_1 + \mathcal{A}_1)^{-1}v$ in Algorithm 4.1.

In the LR-ADI method, we also need the shift parameters that can be computed using the Arnoldi algorithm [Arn51] applied to $\mathcal{E}_1^{-1}\mathcal{A}_1$ and $\mathcal{A}_1^{-1}\mathcal{E}_1$. Therefore, we have to compute the products $\mathcal{E}_1^{-1}\mathcal{A}_1 v$ and $\mathcal{A}_1^{-1}\mathcal{E}_1 v$ for different vectors v. In the following, we present an efficient way to do this without computing the matrices $\mathcal{E}_1$ and $\mathcal{A}_1$ explicitly and without finding the matrix $\mathcal{Y}$.

Algorithm 4.2: Computation of $\mathcal{E}_1^{-1}\mathcal{A}_1 v$

Input : $\mathcal{M}_{11}, \mathcal{K}_{11} \in \mathbb{R}^{n_1\times n_1}$, $\mathcal{K}_{12} \in \mathbb{R}^{n_1\times n_2}$, $\mathcal{K}_{21} \in \mathbb{R}^{n_2\times n_1}$, $\mathcal{K}_{22} \in \mathbb{R}^{n_2\times n_2}$, $\mathcal{X}_1 \in \mathbb{R}^{n_1\times m}$, $\mathcal{X}_2 \in \mathbb{R}^{n_2\times m}$, $\mathcal{R} \in \mathbb{R}^{m\times m}$, $v = [v_1^T, v_2^T]^T \in \mathbb{R}^{n_1+m}$

Output: $z = \mathcal{E}_1^{-1}\mathcal{A}_1 v$ with $\mathcal{E}_1$ and $\mathcal{A}_1$ as in (4.4.1)

1 Compute $\hat{v}_2 = \mathcal{X}_2(\mathcal{X}_2^T\mathcal{X}_2)^{-\frac{1}{2}}v_2$.

2 Solve the linear system

$$\begin{bmatrix} \mathcal{K}_{22} & \mathcal{X}_2 \\ \mathcal{X}_2^T & 0 \end{bmatrix}\begin{bmatrix} z_1 \\ z_2 \end{bmatrix} = \begin{bmatrix} \mathcal{K}_{21}v_1 + \mathcal{K}_{22}\hat{v}_2 \\ 0 \end{bmatrix}.$$

/* $z_1 = \mathcal{Y}(\mathcal{Y}^T\mathcal{K}_{22}\mathcal{Y})^{-1}\mathcal{Y}^T\begin{bmatrix}\mathcal{K}_{21} & \mathcal{K}_{22}\mathcal{Z}\end{bmatrix}v$ */

3 Compute

$$\begin{bmatrix} \hat{z}_1 \\ \hat{z}_2 \end{bmatrix} = -\begin{bmatrix} \mathcal{K}_{11} & \mathcal{K}_{12} \\ \mathcal{K}_{21} & \mathcal{K}_{22} \end{bmatrix}\begin{bmatrix} v_1 \\ \hat{v}_2 \end{bmatrix} + \begin{bmatrix} \mathcal{K}_{12} \\ \mathcal{K}_{22} \end{bmatrix} z_1.$$

4 Compute $w_2 = (\mathcal{X}_2^T\mathcal{X}_2)^{-1}\mathcal{X}_2^T\hat{z}_2$.

5 Solve the linear system $\mathcal{M}_{11}w_1 = \hat{z}_1 - \mathcal{X}_1 w_2$.

/* See the inverse of $\mathcal{E}_1$ in Theorem 4.6 */

6 Compute

$$z = \begin{bmatrix} w_1 \\ -(\mathcal{X}_2^T\mathcal{X}_2)^{-\frac{1}{2}}(\mathcal{X}_1^T w_1 - \mathcal{R}w_2) \end{bmatrix}.$$

Lemma 4.16. *Let $\mathcal{K}_{22} \in \mathbb{R}^{n_2\times n_2}$, $w \in \mathbb{R}^{n_2}$ and let $\mathcal{X}_2 \in \mathbb{R}^{n_2\times m}$ be of full column rank. Let the columns of $\mathcal{Y}$ form a basis of $\ker(\mathcal{X}_2^T)$. If $\left[z_1^T, z_2^T\right]^T$ solves*

$$\begin{bmatrix} \mathcal{K}_{22} & \mathcal{X}_2 \\ \mathcal{X}_2^T & 0 \end{bmatrix}\begin{bmatrix} z_1 \\ z_2 \end{bmatrix} = \begin{bmatrix} w \\ 0 \end{bmatrix}, \tag{4.4.10}$$

then $z_1 = \mathcal{Y}(\mathcal{Y}^T\mathcal{K}_{22}\mathcal{Y})^{-1}\mathcal{Y}^T w$.

Proof. The second equation in (4.4.10) implies that z_1 lies in $\ker(\mathcal{X}_2^T) = \operatorname{im}(\mathcal{Y})$, i.e., there exists $\hat{z}_1 \in \mathbb{R}^{n_2-m}$ such that $z_1 = \mathcal{Y}\hat{z}_1$. Substituting this vector into the first equation in (4.4.10) and multiplying it from the left with $\mathcal{Y}^T$ gives the term $\hat{z}_1 = (\mathcal{Y}^T\mathcal{K}_{22}\mathcal{Y})^{-1}\mathcal{Y}^T w$. Then $z_1 = \mathcal{Y}\hat{z}_1 = \mathcal{Y}(\mathcal{Y}^T\mathcal{K}_{22}\mathcal{Y})^{-1}\mathcal{Y}^T w$. □

Taking into account the definition of $\mathcal{A}_1$ in (4.4.1), the inverse of $\mathcal{E}_1$ in Theorem 4.6 and Lemma 4.16, we can compute $\mathcal{E}_1^{-1}\mathcal{A}_1 v$ using Algorithm 4.2.

Next, we consider the computation of $z = \mathcal{A}_1^{-1}\mathcal{E}_1 v$. Since computing $z = \mathcal{A}_1^{-1}\mathcal{E}_1 v$ is equivalent to solving the linear system $\mathcal{A}_1 z = \mathcal{E}_1 v$, where $\mathcal{A}_1$ is the Schur complement of the matrix given in (4.3.27), $z = \left[z_1^T, z_2^T\right]^T$ can be computed by solving the linear system

$$-\mathcal{T}_2^T\begin{bmatrix} \mathcal{K}_{11} & \mathcal{K}_{12} \\ \mathcal{K}_{21} & \mathcal{K}_{22} \end{bmatrix}\mathcal{T}_2\begin{bmatrix} z_1 \\ z_2 \\ z_3 \end{bmatrix} = \begin{bmatrix} \mathcal{E}_1 v \\ 0 \end{bmatrix}$$

Algorithm 4.3: Computation of $\mathcal{A}_1^{-1}\mathcal{E}_1 v$

Input : $\mathcal{M}_{11}, \mathcal{K}_{11} \in \mathbb{R}^{n_1 \times n_1}$, $\mathcal{K}_{12} \in \mathbb{R}^{n_1 \times n_2}$, $\mathcal{K}_{21} \in \mathbb{R}^{n_2 \times n_1}$, $\mathcal{K}_{22} \in \mathbb{R}^{n_2 \times n_2}$, $\mathcal{X}_1 \in \mathbb{R}^{n_1 \times m}$, $\mathcal{X}_2 \in \mathbb{R}^{n_2 \times m}$, $\mathcal{R} \in \mathbb{R}^{m \times m}$, $v = [v_1^T v_2^T]^T \in \mathbb{R}^{n_1+m}$

Output: $z = \mathcal{A}_1^{-1}\mathcal{E}_1 v$ with $\mathcal{E}_1$ and $\mathcal{A}_1$ as in (4.4.1)

1 Compute $w = \mathcal{R}^{-1}(\mathcal{X}_1^T v + (\mathcal{X}_2^T \mathcal{X}_2)^{\frac{1}{2}} v_2)$.

2 Solve the linear system

$$-\begin{bmatrix} \mathcal{K}_{11} & \mathcal{K}_{12} \\ \mathcal{K}_{21} & \mathcal{K}_{22} \end{bmatrix} \begin{bmatrix} \hat{z}_1 \\ \hat{z}_2 \end{bmatrix} = \begin{bmatrix} \mathcal{M}_{11} v_1 + \mathcal{X}_1 w \\ \mathcal{X}_2 w \end{bmatrix}.$$

3 Compute

$$z = \begin{bmatrix} \hat{z}_1 \\ (\mathcal{X}_2^T \mathcal{X}_2)^{-\frac{1}{2}} \mathcal{X}_2^T \hat{z}_2 \end{bmatrix}.$$

---

with the orthogonal matrix $\mathcal{T}_2$ given in (4.3.28). Multiplying this equation from the left with the matrix $\mathcal{T}_2$ and introducing

$$\begin{bmatrix} \hat{z}_1 \\ \hat{z}_2 \end{bmatrix} = \mathcal{T}_2 \begin{bmatrix} z_1 \\ z_2 \\ z_3 \end{bmatrix}, \tag{4.4.11}$$

$$f = \mathcal{T}_2 \begin{bmatrix} \mathcal{E}_1 v \\ 0 \end{bmatrix} = \begin{bmatrix} I & 0 \\ 0 & Z \end{bmatrix} \mathcal{E}_1 v = \begin{bmatrix} \mathcal{M}_{11} v_1 + \mathcal{X}_1 \mathcal{R}^{-1}(\mathcal{X}_1^T v_1 + (\mathcal{X}_2^T \mathcal{X}_2)^{\frac{1}{2}} v_2) \\ \mathcal{X}_2 \mathcal{R}^{-1}(\mathcal{X}_1^T v_1 + (\mathcal{X}_2^T \mathcal{X}_2)^{\frac{1}{2}} v_2) \end{bmatrix},$$

we get the linear system

$$-\begin{bmatrix} \mathcal{K}_{11} & \mathcal{K}_{12} \\ \mathcal{K}_{21} & \mathcal{K}_{22} \end{bmatrix} \begin{bmatrix} \hat{z}_1 \\ \hat{z}_2 \end{bmatrix} = f.$$

Using the orthogonality of $\mathcal{T}_2$, we obtain from (4.4.11) that

$$\begin{bmatrix} z_1 \\ z_2 \end{bmatrix} = \begin{bmatrix} \hat{z}_1 \\ Z^T \hat{z}_2 \end{bmatrix}.$$

The computation of $\mathcal{A}_1^{-1}\mathcal{E}_1 v$ is summarized in Algorithm 4.3.

Finally, we present in Algorithm 4.4 the computation of the reduced-order model using BT. In the first step, we only solve the Lyapunov equation (4.4.3) for the low-rank Cholesky factor $\tilde{Z}_c$ of the controllability Gramian G_c using the LR-ADI method. It gives us the low-rank factor $\tilde{Z}_o = -\mathcal{E}_1^{-1}\mathcal{A}_1 \tilde{Z}_c$ of the controllability Gramian G_o. Then the matrix $\tilde{Z}_o^T \mathcal{E}_1 \tilde{Z}_c = -\tilde{Z}_c^T \mathcal{A}_1 \tilde{Z}_c$ is symmetric and positive definite. Therefore, we compute the EVD

$$-\tilde{Z}_c^T \mathcal{A}_1 \tilde{Z}_c = \begin{bmatrix} U_1 & U_2 \end{bmatrix} \begin{bmatrix} \Lambda_1 & \\ & \Lambda_2 \end{bmatrix} \begin{bmatrix} U_1 & U_2 \end{bmatrix}^T$$

instead of the more expensive SVD. Note that the product $\mathcal{A}_1 \tilde{Z}_c$ can be calculated analogously to the first four steps of Algorithm 4.2. One can find this in Steps 2 - 4

Algorithm 4.4: Balanced truncation for the 2D linear MQS system

Input : $\mathcal{M}_{11}, \mathcal{K}_{11} \in \mathbb{R}^{n_1 \times n_1}$, $\mathcal{K}_{12} \in \mathbb{R}^{n_1 \times n_2}$, $\mathcal{K}_{21} \in \mathbb{R}^{n_2 \times n_1}$, $\mathcal{K}_{22} \in \mathbb{R}^{n_2 \times n_2}$, $\mathcal{X}_1 \in \mathbb{R}^{n_1 \times m}$, $\mathcal{X}_2 \in \mathbb{R}^{n_2 \times m}$, $\mathcal{R} \in \mathbb{R}^{m \times m}$.

Output: a reduced-order asymptotically stable system $(\tilde{\mathcal{A}}_1, \tilde{\mathcal{B}}_1, \tilde{\mathcal{C}}_1)$.

1 Solve the generalized Lyapunov equation

$$\mathcal{A}_1 G_c \mathcal{E}_1^T + \mathcal{E}_1 G_c \mathcal{A}_1^T = -\mathcal{B}_1 \mathcal{B}_1^T$$

for the low-rank Cholesky factor $\tilde{Z}_c = \left[Z_{c1}^T,\ Z_{c2}^T\right]^T$ of $G_c \approx \tilde{Z}_c \tilde{Z}_c^T$ using the LR-ADI method.

2 Compute $\hat{Z}_{c2} = \mathcal{X}_2(\mathcal{X}_2^T \mathcal{X}_2)^{-\frac{1}{2}} Z_{c2}$.

3 Solve the linear system

$$\begin{bmatrix} \mathcal{K}_{22} & \mathcal{X}_2 \\ \mathcal{X}_2^T & 0 \end{bmatrix} \begin{bmatrix} Z_1 \\ Z_2 \end{bmatrix} = \begin{bmatrix} \mathcal{K}_{21} Z_{c1} + \mathcal{K}_{22} \hat{Z}_{c2} \\ 0 \end{bmatrix}.$$

4 Compute

$$\begin{bmatrix} \hat{Z}_1 \\ \hat{Z}_2 \end{bmatrix} = -\begin{bmatrix} \mathcal{K}_{11} & \mathcal{K}_{12} \\ \mathcal{K}_{21} & \mathcal{K}_{22} \end{bmatrix} \begin{bmatrix} Z_{c1} \\ \hat{Z}_{c2} \end{bmatrix} + \begin{bmatrix} \mathcal{K}_{12} \\ \mathcal{K}_{22} \end{bmatrix} Z_1.$$

5 Compute the EVD

$$-Z_{c1}^T \hat{Z}_1 - \hat{Z}_{c2}^T \hat{Z}_2 = \begin{bmatrix} U_1 & U_2 \end{bmatrix} \begin{bmatrix} \Lambda_1 & 0 \\ 0 & \Lambda_2 \end{bmatrix} \begin{bmatrix} U_1 & U_2 \end{bmatrix}^T,$$

where $\Lambda_1 \in \mathbb{R}^{\eta \times \eta}$ contains the dominant eigenvalues.

6 Compute $\tilde{\mathcal{C}}_1 = -(\mathcal{X}_2^T \mathcal{X}_2)^{-1} \mathcal{X}_2^T \hat{Z}_2 U_1 \Lambda_1^{-\frac{1}{2}}$ and $\tilde{\mathcal{B}}_1 = \tilde{\mathcal{C}}_1^T$.

7 Compute $\tilde{\mathcal{A}}_1 = -(\hat{Z}_1 U_1 \Lambda_1^{-\frac{1}{2}} + \mathcal{X}_1 \tilde{\mathcal{C}}_1)^T \mathcal{M}_{11}^{-1} (\hat{Z}_1 U_1 \Lambda_1^{-\frac{1}{2}} + \mathcal{X}_1 \tilde{\mathcal{C}}_1) - \tilde{\mathcal{C}}_1^T \mathcal{R} \tilde{\mathcal{C}}_1$.

followed by the computation of the EVD in Step 5 of Algorithm 4.4. In the last two steps, we calculate the reduced matrices. Applying the Petrov-Galerkin projection with the projection matrices

$$\begin{aligned} V &= \tilde{Z}_c U_1 \Lambda_1^{-\frac{1}{2}}, \\ W &= \tilde{Z}_o^T U_1 \Lambda_1^{-\frac{1}{2}} = -\mathcal{E}_1^{-1}\mathcal{A}_1 \tilde{Z}_c U_1 \Lambda_1^{-\frac{1}{2}} = -\mathcal{E}_1^{-1}\mathcal{A}_1 V, \end{aligned}$$

we obtain the reduced-order system (4.4.2) with the system matrices

$$\begin{aligned} \tilde{\mathcal{E}}_1 &= W^T \mathcal{E}_1 V = -V^T \mathcal{A}_1 \mathcal{E}_1^{-1} \mathcal{E}_1 V = I, \\ \tilde{\mathcal{A}}_1 &= W^T \mathcal{A}_1 V = -\Lambda_1^{-\frac{1}{2}} U_1^T \tilde{Z}_c^T \mathcal{A}_1 \mathcal{E}_1^{-1} \mathcal{A}_1 \tilde{Z}_c U_1 \Lambda_1^{-\frac{1}{2}}, \\ \tilde{\mathcal{B}}_1 &= W^T \mathcal{B}_1 = -\Lambda_1^{-\frac{1}{2}} U_1^T \tilde{Z}_c^T \mathcal{A}_1 \mathcal{E}_1^{-1} \mathcal{B}_1, \\ \tilde{\mathcal{C}}_1 &= \mathcal{C}_1 V = -\mathcal{B}_1^T \mathcal{E}_1^{-1} \mathcal{A}_1 \tilde{Z}_c U_1 \Lambda_1^{-\frac{1}{2}} = \mathcal{B}_1^T W = \tilde{\mathcal{B}}_1^T. \end{aligned}$$

We now use the matrices $\hat{Z}_1$ and $\hat{Z}_2$ from Step 4 of Algorithm 4.4 and the low-rank Cholesky factor $\tilde{Z}_c$ to calculate

$$\begin{aligned} \mathcal{A}_1 \tilde{Z}_c &= \begin{bmatrix} \hat{Z}_1 \\ (\mathcal{X}_2^T \mathcal{X}_2)^{-\frac{1}{2}} \mathcal{X}_2^T \hat{Z}_2 \end{bmatrix}, \\ \mathcal{E}_1^{-1} \mathcal{A}_1 \tilde{Z}_c &= \begin{bmatrix} W_1 \\ -(\mathcal{X}_2^T \mathcal{X}_2)^{-\frac{1}{2}} (\mathcal{X}_1^T W_1 - \mathcal{R} W_2) \end{bmatrix}, \end{aligned}$$

where $W_2 = (\mathcal{X}_2^T \mathcal{X}_2)^{-1} \mathcal{X}_2^T \hat{Z}_2$ and W_1 solves $\mathcal{M}_{11} W_1 = \hat{Z}_1 - \mathcal{X}_1 W_2$. Then using the block structure of $\mathcal{A}_1$ and $\mathcal{B}_1$, we obtain

$$\begin{aligned} \tilde{\mathcal{A}}_1 &= -\Lambda_1^{-\frac{1}{2}} U_1^T \begin{bmatrix} W_1 \\ -(\mathcal{X}_2^T \mathcal{X}_2)^{-\frac{1}{2}} (\mathcal{X}_1^T W_1 - \mathcal{R} W_2) \end{bmatrix}^T \begin{bmatrix} \hat{Z}_1 \\ (\mathcal{X}_2^T \mathcal{X}_2)^{-\frac{1}{2}} \mathcal{X}_2^T \hat{Z}_2 \end{bmatrix} U_1 \Lambda_1^{-\frac{1}{2}} \\ &= -\Lambda_1^{-\frac{1}{2}} U_1^T (W_1^T \hat{Z}_1 - (\mathcal{X}_1^T W_1 - \mathcal{R} W_2)^T (\mathcal{X}_2^T \mathcal{X}_2)^{-1} \mathcal{X}_2^T \hat{Z}_2) U_1 \Lambda_1^{-\frac{1}{2}} \\ &= \Lambda_1^{-\frac{1}{2}} U_1^T (-W_1^T \hat{Z}_1 + W_1^T \mathcal{X}_1 (\mathcal{X}_2^T \mathcal{X}_2)^{-1} \mathcal{X}_2^T \hat{Z}_2 - W_2^T \mathcal{R} (\mathcal{X}_2^T \mathcal{X}_2)^{-1} \mathcal{X}_2^T \hat{Z}_2) U_1 \Lambda_1^{-\frac{1}{2}} \\ &= \Lambda_1^{-\frac{1}{2}} U_1^T (-W_1^T (\hat{Z}_1 - \mathcal{X}_1 W_2) - W_2^T \mathcal{R} W_2) U_1 \Lambda_1^{-\frac{1}{2}} \\ &= -\Lambda_1^{-\frac{1}{2}} U_1^T ((\hat{Z}_1 - \mathcal{X}_1 W_2)^T \mathcal{M}_{11}^{-1} (\hat{Z}_1 - \mathcal{X}_1 W_2) + W_2^T \mathcal{R} W_2) U_1 \Lambda_1^{-\frac{1}{2}}, \end{aligned}$$

and

$$\begin{aligned} \tilde{\mathcal{C}}_1 &= -\mathcal{B}_1^T \mathcal{E}_1^{-1} \mathcal{A}_1 \tilde{Z}_c U_1 \Lambda_1^{-\frac{1}{2}} \\ &= -\mathcal{R}^{-1} \begin{bmatrix} \mathcal{X}_1 \\ \mathcal{Z}^T \mathcal{X}_2 \end{bmatrix}^T \begin{bmatrix} W_1 \\ -(\mathcal{X}_2^T \mathcal{X}_2)^{-\frac{1}{2}} (\mathcal{X}_1^T W_1 - \mathcal{R} W_2) \end{bmatrix} U_1 \Lambda_1^{-\frac{1}{2}} \\ &= -\mathcal{R}^{-1} (\mathcal{X}_1^T W_1 - \mathcal{X}_1^T W_1 + \mathcal{R} W_2) U_1 \Lambda_1^{-\frac{1}{2}} = -W_2 U_1 \Lambda_1^{-\frac{1}{2}}. \end{aligned}$$

Inserting the definition of W_2 and using $\tilde{\mathcal{C}}_1$ in $\tilde{\mathcal{A}}_1$ gives Steps 6 and 7 in Algorithm 4.4.

It should be noted that Lyapunov-based BT preserves stability, but it, in general, does not guarantee the preservation of passivity in the reduced-order model. Fortunately, due to the special structure of the system matrices in (4.4.1), we can show that the reduced-order model (4.4.2) is passive. Indeed, the reduced-order matrices satisfy

$$\tilde{\mathcal{E}}_1 = \tilde{\mathcal{E}}_1^T > 0, \quad \tilde{\mathcal{A}}_1 = \tilde{\mathcal{A}}_1^T < 0, \quad \tilde{\mathcal{B}}_1 = \tilde{\mathcal{C}}_1^T.$$

Then by Theorem 2.15, system (4.4.2) is io-passive. Furthermore, the controllability Gramian of (4.4.2) is given by $\Lambda_1 > 0$ implying that system (4.4.2) is controllable. Therefore, by Remark 2.17 system (4.4.2) is passive.

4.5 Model reduction for 2D nonlinear MQS systems

For model reduction of the nonlinear MQS system (4.3.5), we apply POD as introduced in Section 2.4.2. To this end, we construct a snapshot matrix

$$\mathcal{X} = \begin{bmatrix} x(t_1), \ldots, x(t_{n_s}) \end{bmatrix}$$

and compute the SVD

$$\mathcal{X} = \begin{bmatrix} U_1 & U_0 \end{bmatrix} \begin{bmatrix} \Sigma_1 & \\ & \Sigma_0 \end{bmatrix} \begin{bmatrix} V_1 & V_2 \end{bmatrix}^T,$$

where $\Sigma_1 \in \mathbb{R}^{\eta\times\eta}$ contains the dominant singular values of $\mathcal{X}$. The reduced-order model can then be determined by projecting

$$\begin{aligned} \tilde{\mathcal{E}}\dot{\tilde{x}} &= \tilde{\mathcal{A}}(\tilde{x})\tilde{x} + \tilde{\mathcal{B}}u, \\ \tilde{y} &= \tilde{\mathcal{C}}\tilde{x} \end{aligned} \tag{4.5.1}$$

with $\tilde{x} \in \mathbb{R}^\eta$, $\tilde{\mathcal{E}} = U_1^T \mathcal{E} U_1$, $\tilde{\mathcal{A}}(\tilde{x}) = U_1^T \mathcal{A}(U_1\tilde{x})U_1$, $\tilde{\mathcal{B}} = U_1\mathcal{B}$ and $\tilde{\mathcal{C}} = \mathcal{C}U_1$. This naive approach has several disadvantages. First note that the algebraic and differential components of the state are mixed in the reduced order model (4.5.1). Secondly, as it has already been mentioned in Section 2.4.1 the reduction of the algebraic equations and states can lead to physically meaningless results.

The idea for our model reduction approach is to use the ODE formulation (4.3.15) introduced in Section 4.3.3 and reduce only the first component a_1 of the state $x_1 = \begin{bmatrix} a_1^T & a_{21}^T \end{bmatrix}^T$, since $a_{21} \in \mathbb{R}^m$ and m is assumed to be small. The snapshot matrix $\mathcal{X}_{a_1}$ can be extracted from $\mathcal{X}$ using the transformation matrix $\mathcal{T}$ given in (4.3.16) by

$$\mathcal{T}\mathcal{X} = \begin{bmatrix} \mathcal{X}_{a_1} \\ \mathcal{X}_{a_{21}} \\ \mathcal{X}_{a_{22}} \\ \mathcal{X}_{\iota} \end{bmatrix}. \tag{4.5.2}$$

Compute now the SVD

$$\mathcal{X}_{a_1} = \begin{bmatrix} U_{a_1} & \hat{U}_{a_1} \end{bmatrix} \begin{bmatrix} \Sigma_{a_1} & \\ & \hat{\Sigma}_{a_1} \end{bmatrix} \begin{bmatrix} V_{a_1} & \hat{V}_{a_1} \end{bmatrix}^T. \tag{4.5.3}$$

Then the reduced-order model is given by

$$\begin{aligned}\tilde{\mathcal{E}}_1\dot{\tilde{x}}_1 &= \tilde{\mathcal{A}}_1(\tilde{x}_1)\tilde{x}_1 + \tilde{\mathcal{B}}_1 u,\\ \tilde{y} &= \tilde{\mathcal{C}}_1\tilde{x}_1,\end{aligned} \tag{4.5.4}$$

with the reduced matrices

$$\tilde{\mathcal{E}}_1 = U^T\mathcal{E}_1 U, \quad \tilde{\mathcal{A}}_1(\tilde{x}_1) = U^T\mathcal{A}_1(U\tilde{x}_1)U, \quad \tilde{\mathcal{B}}_1 = U^T\mathcal{B}_1, \quad \tilde{\mathcal{C}}_1 = \mathcal{C}_1 U, \tag{4.5.5}$$

and the projection matrix

$$U = \begin{bmatrix} U_{a_1} & 0 \\ 0 & I_m \end{bmatrix}. \tag{4.5.6}$$

This model can be computed as in the linear case without calculating the matrices $\mathcal{E}_1$, $\mathcal{A}_1$, $\mathcal{B}_1$ and $\mathcal{C}_1$ explicitly.

In principle, there are several approaches for computing the reduced-order model for the DAE system (4.3.5). Above, we transformed first (4.3.5) into the ODE (4.3.15) and then applied MOR to (4.3.15). This approach is referred to as *first-transform-then-reduce*. The transformation of (4.3.5) into the ODE (4.3.15) with the system matrices as in (4.3.20) and (4.3.21) can also be obtained by projection

$$\mathcal{E}_1 = \mathcal{T}_{l,n_1}\mathcal{E}\mathcal{T}_{r,n_1}, \quad \mathcal{A}_1(x_1) = \mathcal{T}_{l,n_1}\mathcal{A}(\mathcal{T}_{r,n_1}x_1)\mathcal{T}_{r,n_1}, \quad \mathcal{B}_1 = \mathcal{T}_{l,n_1}\mathcal{B}, \quad \mathcal{C}_1 = \mathcal{C}\mathcal{T}_{r,n_1}$$

with the left and right projection matrices given by

$$\mathcal{T}_{l,n_1} = \begin{bmatrix} I_{n_1} & -\mathcal{K}_{12}\mathcal{Y}(\mathcal{Y}^T\mathcal{K}_{22}\mathcal{Y})^{-1}\mathcal{Y}^T & \mathcal{X}_1\mathcal{R}^{-1} \\ 0 & \mathcal{Z}^T - \mathcal{Z}^T\mathcal{K}_{22}\mathcal{Y}(\mathcal{Y}^T\mathcal{K}_{22}\mathcal{Y})^{-1}\mathcal{Y}^T & \mathcal{Z}^T\mathcal{X}_2\mathcal{R}^{-1} \end{bmatrix}, \tag{4.5.7}$$

$$\mathcal{T}_{r,n_1} = \begin{bmatrix} I_{n_1} & 0 & 0 \\ 0 & I_{n_2} & 0 \\ 0 & 0 & (\mathcal{X}_2^T\mathcal{X}_2)^{-1}\mathcal{X}_2^T(I - \mathcal{K}_{22}\mathcal{Y}(\mathcal{Y}^T\mathcal{K}_{22}\mathcal{Y})^{-1}\mathcal{Y}^T) \end{bmatrix} \begin{bmatrix} I_{n_1} & 0 \\ 0 & \mathcal{Z} \\ \mathcal{K}_{21} & \mathcal{K}_{22}\mathcal{Z} \end{bmatrix}, \tag{4.5.8}$$

respectively. Then the reduced-order model (4.5.4) is determined by projection (4.5.5) with the projection matrix U as in (4.5.6).

Alternatively, we can first compute the reduced-order DAE system (4.5.1) by projection

$$\tilde{\mathcal{E}} = \tilde{U}^T\mathcal{E}\tilde{U}, \quad \tilde{\mathcal{A}}(\tilde{x}) = \tilde{U}^T\mathcal{A}(\tilde{U}\tilde{x})\tilde{U}, \quad \tilde{\mathcal{B}} = \tilde{U}^T\mathcal{B}, \quad \tilde{\mathcal{C}} = \mathcal{C}\tilde{U}, \tag{4.5.9}$$

with the projection matrix

$$\tilde{U} = \begin{bmatrix} U_{a_1} & & \\ & I_{n_2} & \\ & & I_m \end{bmatrix}$$

and then transform the resulting DAE into the ODE system (4.3.15) with the system matrices

$$\tilde{\mathcal{E}}_1 = \mathcal{T}_{l,\eta}\tilde{\mathcal{E}}\mathcal{T}_{r,\eta}, \quad \tilde{\mathcal{A}}_1(\tilde{x}_1) = \mathcal{T}_{l,\eta}\tilde{\mathcal{A}}(\mathcal{T}_{r,\eta}\tilde{x}_1)\mathcal{T}_{r,\eta}, \quad \tilde{\mathcal{B}}_1 = \mathcal{T}_{l,\eta}\tilde{\mathcal{B}}, \quad \tilde{\mathcal{C}}_1 = \tilde{\mathcal{C}}\mathcal{T}_{l,\eta}, \tag{4.5.10}$$

with

$$\mathcal{T}_{l,\eta} = \begin{bmatrix} I_\eta & -U_{a_1}^T \mathcal{K}_{12}\mathcal{Y}(\mathcal{Y}^T\mathcal{K}_{22}\mathcal{Y})^{-1}\mathcal{Y}^T & U_{a_1}^T \mathcal{X}_1 \mathcal{R}^{-1} \\ 0 & \mathcal{Z}^T - \mathcal{Z}^T\mathcal{K}_{22}\mathcal{Y}(\mathcal{Y}^T\mathcal{K}_{22}\mathcal{Y})^{-1}\mathcal{Y}^T & \mathcal{Z}^T\mathcal{X}_2\mathcal{R}^{-1} \end{bmatrix}, \tag{4.5.11}$$

$$\mathcal{T}_{r,\eta} = \begin{bmatrix} I_\eta & 0 & 0 \\ 0 & I_{n_2} & 0 \\ 0 & 0 & (\mathcal{X}_2^T\mathcal{X}_2)^{-1}\mathcal{X}_2^T(I - \mathcal{K}_{22}\mathcal{Y}(\mathcal{Y}^T\mathcal{K}_{22}\mathcal{Y})^{-1}\mathcal{Y}) \end{bmatrix} \begin{bmatrix} I_\eta & 0 \\ 0 & \mathcal{Z} \\ \mathcal{K}_{21}U_{a_1} & \mathcal{K}_{22}\mathcal{Z} \end{bmatrix}. \tag{4.5.12}$$

This approach is referred as *first-reduce-then-transform*. Since $U^T\mathcal{T}_{l,n_1} = \mathcal{T}_{l,\eta}\tilde{U}^T$, $\mathcal{T}_{r,n_1}U = \tilde{U}\mathcal{T}_{r,\eta}$ and $\mathcal{A}(x)$ resp. $\mathcal{A}_1(x_1)$ depend only on the first n_1 components of x resp. x_1, which are equal, both MOR approaches are equivalent in the sense that they provide the same reduced-order model. It should be noted that this equivalence holds due to the special structure of the semidiscretized MQS system (4.3.5), (4.3.6). For general DAEs, however, the index reduction and MOR may not commute.

4.5.1 DEIM

In order to speed up the simulation of the reduced-order system (4.5.4), we employ further the DEIM for efficient evaluation of the nonlinearity

$$\tilde{\mathcal{A}}_1(\tilde{x}_1)\tilde{x}_1 = U^T\mathcal{A}_1(U\tilde{x}_1)U\tilde{x}_1. \tag{4.5.13}$$

Using the structure of $\mathcal{A}_1(x_1)$ and $\mathcal{K}(a)$ in (4.3.20b) and (4.3.12), respectively, we separate the nonlinear function $\mathcal{A}_1(x_1)x_1$ into linear and nonlinear parts

$$\mathcal{A}_1(x_1)x_1 = \mathcal{A}_{1l}x_1 + \begin{bmatrix} f_1(a_1) \\ 0 \end{bmatrix},$$

with a constant matrix

$$\mathcal{A}_{1l} = -\begin{bmatrix} \mathcal{K}_{11,l} & \mathcal{K}_{12}\mathcal{Z} \\ \mathcal{Z}^T\mathcal{K}_{21} & \mathcal{Z}^T\mathcal{K}_{22}\mathcal{Z} \end{bmatrix} + \begin{bmatrix} \mathcal{K}_{12} \\ \mathcal{Z}^T\mathcal{K}_{22} \end{bmatrix} \mathcal{Y}(\mathcal{Y}^T\mathcal{K}_{22}\mathcal{Y})^{-1}\mathcal{Y}^T \begin{bmatrix} \mathcal{K}_{21} & \mathcal{K}_{22}\mathcal{Z} \end{bmatrix}$$

and the nonlinear function $f_1(a_1) = \mathcal{K}_{11,n}(a_1)a_1$. Collecting the snapshots of the nonlinearity

$$\mathcal{X}_f = [f_1(a_1(t_1)), \ldots, f_1(a_1(t_{n_s}))], \tag{4.5.14}$$

we compute the SVD

$$\mathcal{X}_f = \begin{bmatrix} U_f & \hat{U}_f \end{bmatrix} \begin{bmatrix} \Sigma_f & \\ & \hat{\Sigma}_f \end{bmatrix} \begin{bmatrix} V_f & \hat{V}_f \end{bmatrix}^T, \tag{4.5.15}$$

where $\begin{bmatrix} U_f & \hat{U}_f \end{bmatrix}$ and $\begin{bmatrix} V_f & \hat{V}_f \end{bmatrix}$ have orthogonal columns and $\Sigma_f \in \mathbb{R}^{\kappa\times\kappa}$ contains the dominant singular values of $\mathcal{X}_f$. Then the nonlinearity (4.5.13) can be approximated as

$$\tilde{\mathcal{A}}_1(\tilde{x}_1)\tilde{x}_1 \approx U^T\mathcal{A}_{1l}U\tilde{x}_1 + \begin{bmatrix} W\mathcal{S}_{\mathcal{K}}^T f_1(U_{a_1}\tilde{a}_1) \\ 0 \end{bmatrix},$$

where $W = U_{a_1}^T U_f(\mathcal{S}_{\mathcal{K}}^T U_f)^{-1}$ and $\mathcal{K} = \{k_1, \ldots, k_\kappa\}$ are obtained by the greedy procedure presented in Algorithm 2.3 or from the QR decomposition of U_f^T as described in [DG16]. The constant matrices $U^T \mathcal{A}_{11} U$ and W can be precomputed and stored in the offline stage, whereas in the online stage, we only evaluate κ components of the function $f_1(U_{a_1}\tilde{a}_1)$ which are given by

$$\big(f_1(U_{a_1}\tilde{a}_1)\big)_k = \int_{\Omega_1} \boldsymbol{\nu}(\cdot, \|\nabla \sum_{i=1}^{n_1} \tilde{\alpha}_i \psi_i\|) \nabla \sum_{i=1}^{n_1} \tilde{\alpha}_i \psi_i \cdot \nabla \psi_k \, d\xi, \tag{4.5.16}$$

for $k \in \mathcal{K}$ and $U_{a_1}\tilde{a}_1 = [\tilde{\alpha}_1, \ldots, \tilde{\alpha}_{n_1}]^T$. We now show how to make this evaluation independent of the dimension n_1. First of all, we note that the integrals in (4.5.16) have only to be computed on $\operatorname{supp}(\psi_k)$, $k \in \mathcal{K}$, which are small subdomains of Ω_1. This means that also $\sum_{i=1}^{n_1} \tilde{\alpha}_i \nabla \psi_i$ has only to be evaluated on $\operatorname{supp}(\psi_k)$. Therefore, we introduce an extended index set

$$\mathcal{K}_{ext,k} = \{i \in \{1, \ldots, n_1\} \quad : \quad \operatorname{int}\big(\operatorname{supp}(\psi_i)\big) \cap \operatorname{int}\big(\operatorname{supp}(\psi_k)\big) \neq \emptyset\}, \tag{4.5.17}$$

where $\operatorname{int}\big(\operatorname{supp}(\psi_k)\big)$ denotes the interior of $\operatorname{supp}(\psi_k)$. For such an index set, we have

$$\sum_{i=1}^{n_1} \tilde{\alpha}_i \psi_i = \sum_{i \in \mathcal{K}_{ext,k}} \tilde{\alpha}_i \psi_i \quad \text{on } \operatorname{supp}(\psi_k).$$

Then the integral (4.5.16) can be simplified to

$$\big(f_1(U_{a_1}\tilde{a}_1)\big)_k = \int\limits_{\operatorname{supp}(\psi_k)} \boldsymbol{\nu}(\cdot, \|\nabla \sum_{i \in \mathcal{K}_{ext,k}} \tilde{\alpha}_i \psi_i\|) \nabla \sum_{i \in \mathcal{K}_{ext,k}} \tilde{\alpha}_i \psi_i \cdot \nabla \psi_k \, d\xi, \quad k \in \mathcal{K}.$$

One can see that to evaluate the function $\mathcal{S}_{\mathcal{K}}^T f_1(U_1 a_1)$, we do not need all components of $U_1 a_1 \in \mathbb{R}^{n_1}$, but rather only those from the index set

$$\mathcal{K}_{ext} = \bigcup_{k \in \mathcal{K}} \mathcal{K}_{ext,k}, \tag{4.5.18}$$

whose number of elements, denoted by $|\mathcal{K}_{ext}|$, is much smaller than n_1. A simple example for construction of the sets $\mathcal{K}$ and $\mathcal{K}_{ext}$ is presented in Figure 4.2(a). We introduce now the function $\hat{\hat{f}}_1 : \mathbb{R}^{|\mathcal{K}_{ext}|} \to \mathbb{R}^{\kappa}$ as

$$\hat{\hat{f}}_1(\mathcal{S}_{\mathcal{K}_{ext}}^T U_{a_1}\tilde{a}_1) = \mathcal{S}_{\mathcal{K}}^T f_1(U_{a_1}\tilde{a}_1), \tag{4.5.19}$$

where $\mathcal{S}_{\mathcal{K}_{ext}}$ is the selector matrix associated with $\mathcal{K}_{ext}$. This function coincides with $\mathcal{S}_{\mathcal{K}}^T f_1$ but unlike $\mathcal{S}_{\mathcal{K}}^T f_1$, it depends only on the selected components of $U_{a_1}\tilde{a}_1$. This means that the DEIM approximation

$$\hat{f}_1(\tilde{a}_1) = W \hat{\hat{f}}_1(\mathcal{S}_{\mathcal{K}_{ext}}^T U_{a_1}\tilde{a}_1)$$

can be calculated independently of the original size n_1.

As a result, we obtain the POD-DEIM reduced model

$$\begin{aligned} \hat{\mathcal{E}}_1 \dot{\hat{x}}_1 &= \hat{\mathcal{A}}_1(\hat{x}_1)\hat{x}_1 + \hat{\mathcal{B}}_1 u, \\ \hat{y} &= \hat{\mathcal{C}}_1 \hat{x}_1, \end{aligned} \tag{4.5.20}$$

where $\hat{x}_1 = \left[\hat{a}_1^T,\, \hat{a}_{21}^T\right]^T \in \mathbb{R}^{\eta+m}$, and

$$\begin{aligned}\hat{\mathcal{E}}_1 = \tilde{\mathcal{E}}_1, \quad \hat{\mathcal{B}}_1 = \tilde{\mathcal{B}}_1, \quad \hat{\mathcal{C}}_1 = \tilde{\mathcal{C}}_1,\\ \hat{\mathcal{A}}_1(\hat{x}_1)\hat{x}_1 = U^T \mathcal{A}_{1l} U \hat{x}_1 + \begin{bmatrix} \hat{f}_1(\hat{a}_1) \\ 0 \end{bmatrix}.\end{aligned} \tag{4.5.21}$$

In Figure 4.1, we present a workflow diagram which contains all systems we are working with starting from the weak formulation and ending with the POD-DEIM reduced model.

4.5.2 Computing the Jacobi matrix

Integrating this system in time using an one-step or multistep method [HNW93], we face with the problem of solving a sequence of systems of nonlinear equations. For this purpose, we employ the Newton iteration which requires the computation of the Jacobi matrix $J_{\hat{f}_1}(\hat{a}_1)$ of the nonlinear function $\hat{f}_1$ at $\hat{a}_1$ given by

$$J_{\hat{f}_1}(\hat{a}_1) = W \mathcal{S}_{\mathcal{K}}^T J_{f_1}(U_{a_1}\hat{a}_1) U_{a_1}, \tag{4.5.22}$$

where $J_{f_1}(U_{a_1}\hat{a}_1)$ is the Jacobi matrix of f_1 at $U_{a_1}\hat{a}_1$. In this section, we present two different approaches for efficient computation of this matrix.

The first approach for the efficient computation of $J_{\hat{f}_1}$ is based on the assumption that the matrix $\mathcal{S}_{\mathcal{K}}^T J_{f_1}(U_{a_1}\hat{a}_1)$ is sparse. In this case, we introduce an index set

$$\mathcal{J} = \{(i,j) \;:\; \left(\mathcal{S}_{\mathcal{K}}^T J_{f_1}(U_{a_1}\hat{a}_1)\right)_{i,j} \neq 0\} \tag{4.5.23}$$

of non-zero entries of this matrix. Let the matrices $\Theta_{(i,j)} \in \mathbb{R}^{\kappa \times n_1}$ have all zero entries except for the (i,j)-th entry being 1. Then $J_{\hat{f}_1}(\hat{a}_1)$ admits an affine representation

$$J_{\hat{f}_1}(\hat{a}_1) = \sum_{(i,j)\in\mathcal{J}} W\Theta_{(i,j)} U_{a_1} \left(\mathcal{S}_{\mathcal{K}}^T J_{f_1}(U_{a_1}\hat{a}_1)\right)_{i,j}, \tag{4.5.24}$$

where the time-independent matrices $W\Theta_{(i,j)}U_{a_1} \in \mathbb{R}^{\eta\times\eta}$ can be precomputed and stored in the offline phase and only a small number of the time-dependent functions $\left(\mathcal{S}_{\mathcal{K}}^T J_{f_1}(U_{a_1}\hat{a}_1)\right)_{(i,j)}$, $(i,j) \in \mathcal{J}$, have to be evaluated in the online phase.

An alternative approach is based on MDEIM as discussed in Section 2.4.3. In this method, we first compute the basis matrices $V_1, \ldots, V_\rho$ as in (2.4.14) form the EVD of the snapshot matrix

$$\mathcal{X}_J = \begin{bmatrix} \langle J_1, J_1\rangle_F & \ldots & \langle J_1, J_{n_s}\rangle_F \\ \vdots & \ddots & \vdots \\ \langle J_{n_s}, J_1\rangle_F & \ldots & \langle J_{n_s}, J_{n_s}\rangle_F \end{bmatrix} \tag{4.5.25}$$

with $J_i = \mathcal{S}_{\mathcal{K}}^T J_{f_1}(a_1(t_i))$, $i = 1, \ldots, n_s$. Then we determine the MDEIM index set $\mathcal{J}$ and the matrix G_ρ using the MDEIM greedy procedure as presented in Algorithm 2.4. Finally, using (4.5.22) we obtain the MDEIM approximation of the Jacobi matrix

$$J_{\hat{f}_1}(\hat{a}_1) \approx \sum_{l=1}^{\rho} W V_l U_{a_1} g_l(\hat{a}_1) \tag{4.5.26}$$

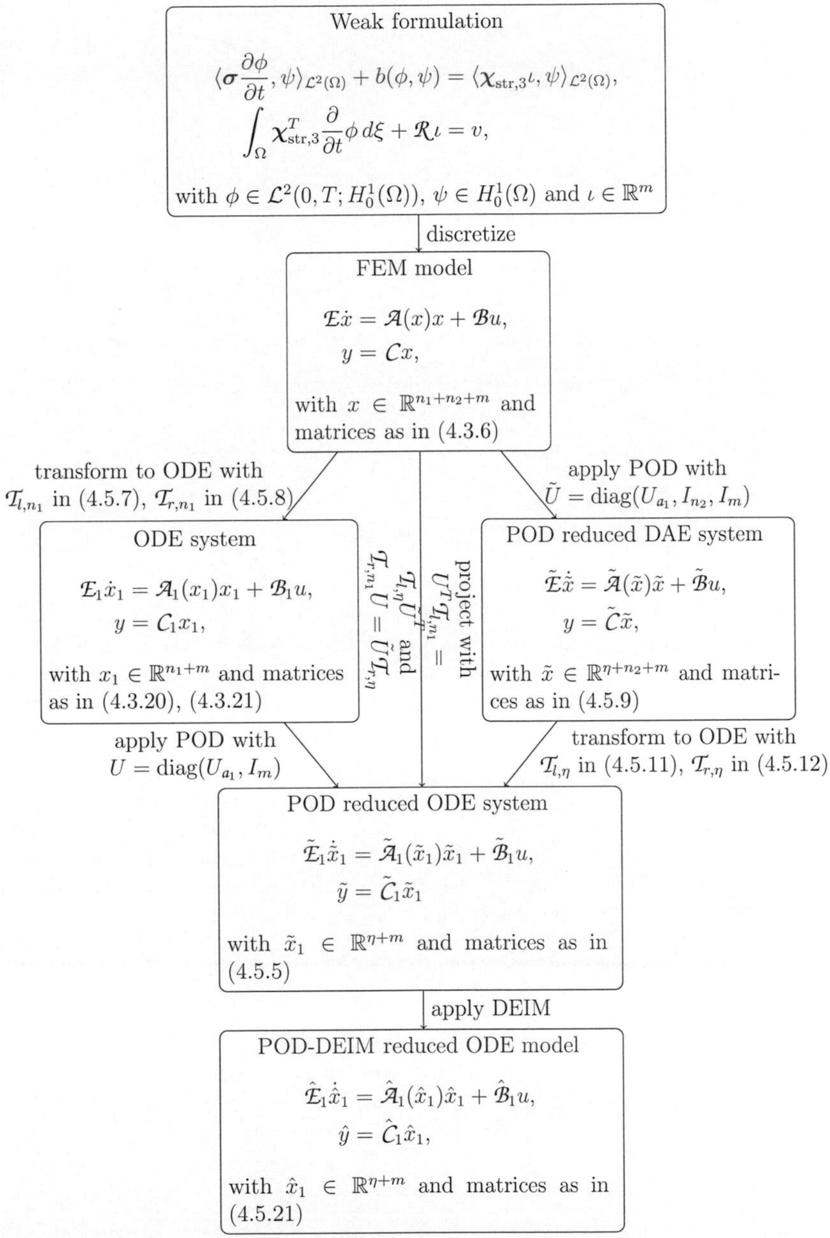

Figure 4.1: Workflow diagram for discretization and model reduction of the 2D nonlinear MQS system

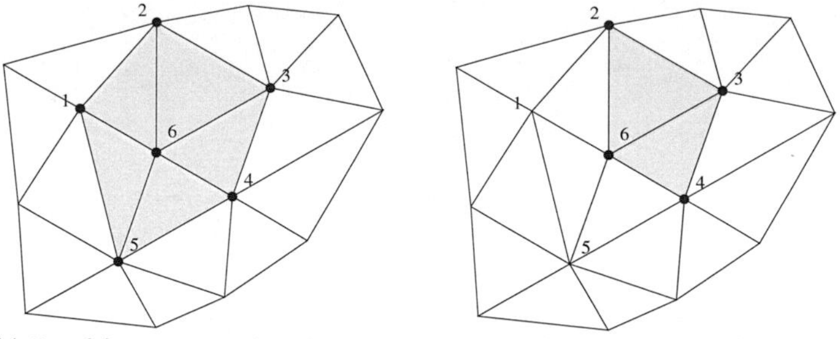

(a) $\mathcal{K} = \{6\}$ and $\mathcal{K}_{ext} = \{1,2,3,4,5,6\}$ for a vector-valued function

(b) $\mathcal{J} = \{(6,3)\}$ and $\mathcal{J}_{ext} = \{2,3,4,6\}$ for a matrix-valued function

Figure 4.2: An example for the (extended) index sets and the integrated domains

with

$$\begin{bmatrix} g_1(\hat{a}_1) \\ \vdots \\ g_\rho(\hat{a}_1) \end{bmatrix} = G_\rho^{-1}\theta(U_{a_1}\hat{a}_1), \quad \theta(U_{a_1}\hat{a}_1) = \begin{bmatrix} (\mathcal{S}_\mathcal{K}^T J_{f_1}(U_{a_1}\hat{a}_1))_{i_1,j_1} \\ \vdots \\ (\mathcal{S}_\mathcal{K}^T J_{f_1}(U_{a_1}\hat{a}_1))_{i_\rho,j_\rho} \end{bmatrix}$$

and $(i_l, j_l) \in \mathcal{J}$, $l = 1, \ldots, \rho$. In both approaches, we have to evaluate only selected components $(\mathcal{S}_\mathcal{K}^T J_{f_1}(U_{a_1}\hat{a}_1))_{i_l,j_l}$ for $(i_l, j_l) \in \mathcal{J}$, where $\mathcal{J}$ is either defined in (4.5.23) or determined by MDEIM. Analogously to the evaluation of $\mathcal{S}_\mathcal{K}^T f_1$ in (4.5.19), thus the components depend only on a few components of $U_{a_1}\hat{a}_1$ determined by an extended index set

$$\mathcal{J}_{ext} = \bigcup_{(i,j)\in\mathcal{J}} \{l \in \{1,\ldots,n_1\} : \mathrm{int}(\mathrm{supp}(\phi_l)) \cap \mathrm{int}(\mathrm{supp}(\phi_i)) \cap \mathrm{int}(\mathrm{supp}(\phi_j)) \neq \emptyset\}. \tag{4.5.27}$$

An exemplary construction of such a set on the integration domain is presented in Figure 4.2(b). To emphasize the dependency on the selected components of $U_{a_1}\hat{a}_1$ we rewrite the sparse representation (4.5.24) as

$$J_{\hat{f}_1}(\hat{a}_1) = \sum_{(i,j)\in\mathcal{J}} W\Theta_{(i,j)}U_{a_1}\left(\mathcal{S}_\mathcal{K}^T J_{f_1}(\mathcal{S}_{\mathcal{J}_{ext}}^T U_{a_1}\hat{a}_1)\right)_{i,j},$$

where $\mathcal{S}_{\mathcal{J}_{ext}}$ denotes the selector matrix associated with $\mathcal{J}_{ext}$. In the MDEIM approach, the approximated Jacobi matrix is obtained as in (4.5.26), where

$$\begin{bmatrix} g_1(\hat{a}_1) \\ \vdots \\ g_\rho(\hat{a}_1) \end{bmatrix} = G_\rho^{-1}\hat{\theta}(\mathcal{S}_{\mathcal{J}_{ext}}^T U_{a_1}\hat{a}_1) \tag{4.5.28}$$

with a new function $\hat{\theta}(\mathcal{S}_{\mathcal{J}_{ext}}^T U_{a_1}\hat{a}_1) = \theta(U_{a_1}\hat{a}_1)$. Thus, in both cases, the approximated Jacobi matrix can be computed independently of the original size n_1.

1. Compute the snapshot matrices $\mathcal{X}_{a_1} = \left[a_1(t_1), \ldots, a_1(t_{n_s})\right]$ and $\mathcal{X}_f$ as in (4.5.14).
2. Construct POD projection matrices $U_{a_1} \in \mathbb{R}^{n_1 \times \eta}$ and $U_f \in \mathbb{R}^{n_1 \times \kappa}$ from the SVDs (4.5.3) and (4.5.15), respectively.
3. Select the index set $\mathcal{K} = \{k_1, \ldots, k_\kappa\}$ using the DEIM greedy procedure in Algorithm 2.3 applied to U_f and construct $\mathcal{K}_{ext,k}$ as in (4.5.17) and $\mathcal{K}_{ext}$ as in (4.5.18).
4. Compute the snapshot matrix $\mathcal{X}_J$ as in (4.5.25) and construct the basis matrices $V_1, \ldots, V_\rho \in \mathbb{R}^{\kappa \times n_1}$ as in (2.4.14) using the EVD (2.4.13).
5. Select the index set $\mathcal{J}$ and the matrix G_ρ using MDEIM Algorithm 2.4 applied to the matrices $V_1, \ldots, V_\rho$ and construct $\mathcal{J}_{ext}$ as in (4.5.27).
6. Compute and store the time-independent matrices
$\tilde{\mathcal{E}}_1 = U^T \mathcal{E}_1 U \in \mathbb{R}^{(\eta+m)\times(\eta+m)}$,
$\tilde{\mathcal{B}}_1 = U^T \mathcal{B}_1 \in \mathbb{R}^{(\eta+m)\times m}$,
$\tilde{\mathcal{C}}_1 = \mathcal{C}_1 U \in \mathbb{R}^{m\times(\eta+m)}$,
$\tilde{\mathcal{A}}_{1l} = U^T \mathcal{A}_{1l} U \in \mathbb{R}^{(\eta+m)\times(\eta+m)}$ with U as in (4.5.6),
$W = U_{a_1}^T U_f (\mathcal{S}_{\mathcal{K}}^T U_f)^{-1} \in \mathbb{R}^{\eta\times\kappa}$ with $\mathcal{S}_{\mathcal{K}} = \left[e_{k_1} \ \ldots \ e_{k_\kappa}\right]$,
$U_{\mathcal{K}_{ext}} = \mathcal{S}_{\mathcal{K}_{ext}}^T U_{a_1} \in \mathbb{R}^{|\mathcal{K}_{ext}|\times\eta}$ with the selector $\mathcal{S}_{\mathcal{K}_{ext}}$ associated with $\mathcal{K}_{ext}$,
$\tilde{V}_l = W V_l U_{a_1} \in \mathbb{R}^{\eta\times\eta}$ for $l = 1, \ldots, \rho$,
$U_{\mathcal{J}_{ext}} = \mathcal{S}_{\mathcal{J}_{ext}}^T U_{a_1} \in \mathbb{R}^{|\mathcal{J}_{ext}|\times\eta}$ with the selector $\mathcal{S}_{\mathcal{J}_{ext}}$ associated with $\mathcal{J}_{ext}$.

Figure 4.3: Offline stage of the POD-DEIM-MDEIM reduction

4.5.3 Online/offline decomposition

As it was mentioned above, the model reduction procedure and simulation of the resulting reduced-order model admit the decomposition into a computationally expensive offline stage and a rapid online stage. The offline stage is presented in Figure 4.3. In order to reduce computational complexity, the snapshot matrices $\mathcal{X}_{a_1}$ and $\mathcal{X}_f$ are generated by solving the DAE (4.3.5). The reduced-order matrices $\tilde{\mathcal{E}}_1$, $\tilde{\mathcal{B}}_1$, $\tilde{\mathcal{C}}_1$ and $\tilde{\mathcal{A}}_{1l}$ are computed as in the linear case without forming the matrices $\mathcal{E}_1$, $\mathcal{B}_1$, $\mathcal{C}_1$ and $\mathcal{A}_{1l}$ explicitly by using Lemma 4.16 and the structure of these matrices. In the online stage, we solve the POD-DEIM reduced-order model (4.5.20). The approximate Jacobi matrix of $\hat{f}(\hat{x}_1) = \hat{\mathcal{A}}_1(\hat{x}_1)\hat{x}_1$ at $\hat{x}_1$ is given by

$$J_{\hat{f}}(\hat{x}_1) = \tilde{\mathcal{A}}_{1l} + \begin{bmatrix} \sum_{l=1}^{\rho} \tilde{V}_l g_l(\hat{a}_1) \\ 0 \end{bmatrix}$$

with g_l given in (4.5.28). If we exploit the sparsity of the Jacobi matrix of f_1 instead of using MDEIM, then Steps 4 and 5 in the offline stage should be replaced by

4'. Construct the index set $\mathcal{J}$ as in (4.5.23) and $\mathcal{J}_{ext}$ as in (4.5.27).

5'. For $(i_l, j_l) \in \mathcal{J}$, set $\tilde{V}_l = \Theta_{(i_l,j_l)} \in \mathbb{R}^{\kappa \times n_1}$, $l = 1, \ldots, \rho$, where $\Theta_{(i_l,j_l)}$ has all zero entries except for the (i_l, j_l)-th entry being 1.

Furthermore, in Step 6, we replace the computation of $\tilde{V}_l$ with

$$\tilde{V}_l = W\Theta_{(i_l,j_l)} U_{a_1} \in \mathbb{R}^{\eta \times \eta} \text{ for } l = 1, \ldots, \rho.$$

Then in the online phase, for solving system (4.5.20), we use the Jacobi matrix given by

$$J_{\hat{f}}(\hat{x}_1) = \tilde{\mathcal{A}}_{1l} + \begin{bmatrix} \sum_{l=1}^{\rho} \tilde{V}_l g_l(\hat{a}_1) \\ 0 \end{bmatrix} \tag{4.5.29}$$

with g_l given in (4.5.28). One can see that in both variants, the online stage computations are independent of the original dimension $n_a + m$ of the problem.

4.5.4 Passivity of the POD reduced system

We now show that passivity is preserved in the POD reduced system (4.5.4), (4.5.5).

Theorem 4.17. *The POD reduced system* (4.5.4), (4.5.5) *is passive.*

Proof. The result can be proved analogously to Theorem 4.11. Using (4.3.26) and the approximation

$$\begin{bmatrix} a_1 \\ a_{21} \end{bmatrix} \approx \begin{bmatrix} U_{a_1} & 0 \\ 0 & I_m \end{bmatrix} \begin{bmatrix} \tilde{a}_1 \\ \tilde{a}_{21} \end{bmatrix},$$

we find

$$\begin{bmatrix} a_1 \\ a_2 \end{bmatrix} = R \begin{bmatrix} a_1 \\ a_{21} \end{bmatrix} \approx \tilde{R} \begin{bmatrix} \tilde{a}_1 \\ \tilde{a}_{21} \end{bmatrix},$$

where

$$\tilde{R} = \begin{bmatrix} U_{a_1} & 0 \\ -\mathcal{Y}(\mathcal{Y}^T \mathcal{K}_{22} \mathcal{Y})^{-1} \mathcal{Y}^T \mathcal{K}_{21} U_{a_1} & (I_{n_2} - \mathcal{Y}(\mathcal{Y}^T \mathcal{K}_{22} \mathcal{Y})^{-1} \mathcal{Y}^T \mathcal{K}_{22}) \mathcal{Z} \end{bmatrix}. \tag{4.5.30}$$

Introducing new basis functions

$$[\tilde{\phi}_1(\xi), \ldots, \tilde{\phi}_{\eta+m}(\xi)] = [\psi_1(\xi), \ldots, \psi_{n_a}(\xi)]\tilde{R},$$

we obtain

$$\sum_{i=1}^{n_a} \alpha_i(t)\psi_i(\xi) \approx \sum_{i=1}^{\eta+m} \tilde{\alpha}_i(t)\tilde{\phi}_i(\xi),$$

where

$$[\tilde{\alpha}_1(t), \ldots, \tilde{\alpha}_\eta(t)]^T = \tilde{a}_1(t)$$

and

$$[\tilde{\alpha}_{\eta+1}(t), \ldots, \tilde{\alpha}_{\eta+m}(t)]^T = \tilde{a}_{21}(t).$$

For $\tilde{x}_1(t) = [\tilde{\alpha}_1(t), \ldots, \tilde{\alpha}_{\eta+m}(t)]^T$, we define a storage function

$$\tilde{S}(\tilde{x}_1(t)) = \int_\Omega \vartheta(\xi, \|\nabla \sum_{i=1}^{\eta+m} \tilde{\alpha}_i(t)\tilde{\phi}_i(\xi)\|^2)\, d\xi,$$

with ϑ as in (4.3.25). This function is nonnegative, since $\boldsymbol{\nu}$ is positive and $\tilde{S}(0) = 0$ due to the definition of ϑ. We calculate

$$\begin{aligned}
\frac{d}{dt}\tilde{S}(\tilde{x}_1(t)) &= \frac{d}{dt}\int_\Omega \vartheta(\xi, \|\nabla \sum_{i=1}^{\eta+m} \tilde{\alpha}_i(t)\tilde{\phi}_i(\xi)\|^2)\,d\xi \\
&= \int_\Omega \frac{\partial}{\partial \varrho}\vartheta(\xi, \|\nabla \sum_{i=1}^{\eta+m} \tilde{\alpha}_i(t)\tilde{\phi}_i(\xi)\|^2)\frac{\partial}{\partial t}\|\nabla \sum_{i=1}^{\eta+m} \tilde{\alpha}_i(t)\tilde{\phi}_i(\xi)\|^2\,d\xi \\
&= \int_\Omega \boldsymbol{\nu}(\xi, \|\nabla \sum_{i=1}^{\eta+m} \tilde{\alpha}_i(t)\tilde{\phi}_i(\xi)\|)\cdot\left(\nabla \sum_{i=1}^{\eta+m} \tilde{\alpha}_i(t)\tilde{\phi}_i(\xi)\right)\cdot\left(\nabla \sum_{i=1}^{\eta+m} \dot{\tilde{\alpha}}_i(t)\tilde{\phi}_i(\xi)\right)d\xi \\
&= \sum_{j=1}^{\eta+m}\tilde{\alpha}_j(t)\sum_{k=1}^{\eta+m}\dot{\tilde{\alpha}}_k(t)\int_\Omega \boldsymbol{\nu}(\xi, \|\nabla \sum_{i=1}^{\eta+m} \tilde{\alpha}_i(t)\tilde{\phi}_i(\xi)\|)\Big(\nabla\tilde{\phi}_j(\xi)\Big)\cdot\Big(\nabla\tilde{\phi}_k(\xi)\Big)\,d\xi \\
&= \dot{\tilde{x}}_1^T(t)\tilde{R}^T\mathcal{K}(\tilde{R}\tilde{x}_1(t))\tilde{R}\tilde{x}_1(t).
\end{aligned}$$

Using the relations

$$\begin{aligned}
\tilde{\mathcal{A}}_1(\tilde{x}_1) &= -\tilde{R}^T\mathcal{K}(\tilde{R}\tilde{x}_1)\tilde{R}, \\
\tilde{\mathcal{E}}_1 &= \tilde{R}^T(\mathcal{M} + X\mathcal{R}^{-1}X^T)\tilde{R} = \begin{bmatrix} U_{a_1}^T\mathcal{M}_{11}U_{a_1} & 0 \\ 0 & 0\end{bmatrix} + \tilde{\mathcal{B}}_1\mathcal{R}\tilde{\mathcal{B}}_1^T,
\end{aligned}$$

and the state equation (4.5.4), we can continue

$$\begin{aligned}
\frac{d}{dt}\tilde{S}(\tilde{x}_1(t)) &= -\dot{\tilde{x}}_1^T(t)\tilde{\mathcal{A}}_1(\tilde{x}_1(t))\tilde{x}_1(t) \\
&= -\dot{\tilde{x}}_1^T(t)\tilde{\mathcal{E}}_1\dot{\tilde{x}}_1(t) + \dot{\tilde{x}}_1^T(t)\tilde{\mathcal{B}}_1u(t) \\
&= -\dot{\tilde{a}}_1^T(t)U_{a_1}^T\mathcal{M}_{11}U_{a_1}\dot{\tilde{a}}_1(t) - \dot{\tilde{x}}_1^T(t)\tilde{\mathcal{B}}_1\mathcal{R}\tilde{\mathcal{B}}_1^T\dot{\tilde{x}}_1(t) + \dot{\tilde{x}}_1^T(t)\tilde{\mathcal{B}}_1u(t) \\
&= -\dot{\tilde{a}}_1^T(t)U_{a_1}^T\mathcal{M}_{11}U_{a_1}\dot{\tilde{a}}_1(t) + \dot{\tilde{x}}_1^T(t)\tilde{\mathcal{B}}_1(u(t) - \mathcal{R}\tilde{\mathcal{B}}_1^T\dot{\tilde{x}}_1(t)).
\end{aligned}$$

Since the matrix $\mathcal{M}_{11}$ is positive definite, the first summand is negative. Furthermore, we use the output equation $\tilde{y} = -\tilde{\mathcal{B}}_1^T\dot{\tilde{x}}_1 + \mathcal{R}^{-1}u$ twice and obtain

$$\begin{aligned}
\frac{d}{dt}\tilde{S}(\tilde{x}_1(t)) &\leqslant (\mathcal{R}^{-1}u(t) - \tilde{y}(t))^T\mathcal{R}(\mathcal{R}^{-1}u(t) + \tilde{y}(t) - \mathcal{R}^{-1}u(t)) \\
&= -\tilde{y}^T(t)\mathcal{R}\tilde{y}(t) + \tilde{y}^T(t)u(t) \\
&\leqslant \tilde{y}^T(t)u(t).
\end{aligned}$$

Integrating this inequality on $[0, T]$, we get the passivation inequality

$$\tilde{S}(\tilde{x}_1(T)) - \tilde{S}(\tilde{x}_1(0)) \leqslant \int_0^T \tilde{y}^T(t)u(t)\,dt$$

which implies the passivity of the reduced-order model (4.5.4). □

Overall, we have shown that the variational MQS problem (4.3.3), (4.3.2b), (4.3.2d), the semidiscretized MQS equation (4.3.5), (4.3.6) and the POD-reduced model

(4.5.4), (4.5.5) are passive. The next step is to verify the passivity of the POD-DEIM reduced model (4.5.20). In [AH17, PM16], a symplectic DEIM was developed for Hamiltonian systems which preserves passivity. Since system (4.3.5) does not have the Hamiltonian form, we cannot use this method here. In the next section, we present a perturbation-based method to enforce passivity for the POD-DEIM reduced model (4.5.20).

4.5.5 Enforcing passivity for the POD-DEIM reduced model

Since DEIM does not preserve the symmetric structure that was used for the construction of the storage functions above, we cannot use our approach. This makes it difficult to verify passivity of the POD-DEIM reduced model (4.5.20). Therefore, we try to ensure the io-passivity for a slightly modified system. Our goal is now to find a scalar function δ such that the perturbed system

$$\begin{aligned} \hat{\mathcal{E}}_1\dot{\hat{x}}_1 &= \hat{\mathcal{A}}_1(\hat{x}_1)\hat{x}_1 + \hat{\mathcal{B}}_1 u, \\ y_\delta &= \hat{\mathcal{C}}_1\hat{x}_1 + \delta u \end{aligned} \tag{4.5.31}$$

is io-passive and the output error $\|\hat{y}-y_\delta\|$ is small. For this purpose, we consider the POD reduced system (4.5.4) with the state $\tilde{x}_1$ and the POD-DEIM reduced system (4.5.20) with the state $\hat{x}_1$. Then for the state error $\varepsilon_1 = \tilde{x}_1 - \hat{x}_1$ and $\hat{\mathcal{C}}_1 = \tilde{\mathcal{C}}_1$, we have

$$\begin{aligned} \int_0^T y_\delta^T(t)u(t)\,dt &= \int_0^T \Big(\hat{\mathcal{C}}_1\hat{x}_1(t) + \delta(t)u(t)\Big)^T u(t)\,dt \\ &= \int_0^T \Big(\tilde{\mathcal{C}}_1\tilde{x}_1(t) - \tilde{\mathcal{C}}_1\varepsilon_1(t) + \delta(t)u(t)\Big)^T u(t)\,dt \\ &\geqslant \int_0^T \tilde{y}^T(t)u(t)\,dt + \int_0^T \delta(t)\|u(t)\|^2\,dt - \int_0^T \|\tilde{\mathcal{C}}_1\|_2\|\varepsilon_1(t)\|\|u(t)\|dt. \end{aligned} \tag{4.5.32}$$

Since the POD reduced system (4.5.4) is io-passive, the first integral in (4.5.32) is nonnegative. Furthermore, by choosing

$$\delta(t) \geqslant \begin{cases} \frac{\|\tilde{\mathcal{C}}_1\|_2\|\varepsilon_1(t)\|}{\|u(t)\|}, & \text{if } \|u(t)\| \neq 0, \\ 0, & \text{if } \|u(t)\| = 0, \end{cases}$$

we get $\delta(t)\|u(t)\|^2 - \|\tilde{\mathcal{C}}_1\|_2\|\varepsilon_1(t)\|\|u(t)\| \geqslant 0$ for all $t \in [0,T]$. In this case, we obtain

$$\int_0^T y_\delta^T(t)u(t)\,dt \geqslant 0$$

and hence, the perturbed system (4.5.31) is io-passive.

The computation of $\delta(t)$ relies on the DEIM error $\varepsilon_1(t)$ which is not readily available. Therefore, we aim to get a computable bound $\|\varepsilon_1(t)\| \leq \epsilon(t)$ which would allow us to easy determine $\delta(t)$ as

$$\delta(t) = \begin{cases} \frac{\|\tilde{\mathcal{C}}_1\|_2\epsilon(t)}{\|u(t)\|}, & \text{if } \|u(t)\| \neq 0, \\ 0, & \text{if } \|u(t)\| = 0, \end{cases} \tag{4.5.33}$$

and, as a consequence, also the output errors

$$\|\tilde{y}(t)-\hat{y}(t)\| = \|\tilde{C}_1\tilde{x}_1(t)-\tilde{C}_1\hat{x}_1(t)\| \leqslant \|\tilde{C}_1\|_2\|\varepsilon_1(t)\| \leqslant \|\tilde{C}_1\|_2\epsilon(t), \qquad (4.5.34)$$
$$\|\hat{y}(t)-y_\delta(t)\| = \|\delta(t)u(t)\| = \|\tilde{C}_1\|_2\epsilon(t).$$

Note that $\delta(t)$ in (4.5.33) is unbounded if $u(t)$ takes zero values. Nevertheless, the perturbation $\delta(t)u(t)$ remains bounded provided $\epsilon(t)$ is bounded. In order to derive a bound on $\|\varepsilon_1(t)\|$, we make use of a logarithmic Lipschitz constant $L_2[\tilde{f}]$ for a nonlinear function $\tilde{f}(\tilde{x}_1) = \tilde{\mathcal{A}}_1(\tilde{x}_1)\tilde{x}_1$. The following theorem provides a bound on the DEIM state error ε_1.

Theorem 4.18. *Consider the POD reduced system* (4.5.4) *with the state* $\tilde{x}_1$ *and the POD-DEIM reduced system* (4.5.20) *with the state* $\hat{x}_1$. *Then the error* $\varepsilon_1 = \tilde{x}_1 - \hat{x}_1$ *can be estimated as*

$$\|\varepsilon_1(t)\| \leqslant \frac{\beta}{L_2[\tilde{f}]}\left(e^{\frac{L_2[\tilde{f}]}{\lambda_{\min}(\tilde{\mathcal{E}}_1)}t}-1\right),$$

where $\beta = \|(\mathcal{S}_{\mathcal{K}}^T U_f)^{-1}\|_2 \sum_{i=\eta+1}^{n_s}\sigma_i(\mathcal{X}_f)$ *with the singular values* $\sigma_i(\mathcal{X}_f)$ *of the DEIM snapshot matrix* $\mathcal{X}_f$, *and* $L_2[\tilde{f}]$ *is the logarithmic Lipschitz constant of* $\tilde{f}(\tilde{x}_1) = \tilde{\mathcal{A}}(\tilde{x}_1)\tilde{x}_1$.

Proof. Subtracting the POD-DEIM system (4.5.20) from the POD system (4.5.4) and taking into account that $\hat{\mathcal{E}}_1 = \tilde{\mathcal{E}}_1$ and $\hat{\mathcal{B}}_1 = \tilde{\mathcal{B}}_1$, we obtain the following system for the error

$$\tilde{\mathcal{E}}_1\dot{\varepsilon}_1 = \tilde{\mathcal{A}}_{1l}\varepsilon_1 + \begin{bmatrix} U_{a_1}^T f_1(U_{a_1}\tilde{a}_1) - \hat{f}_1(\hat{a}_1) \\ 0 \end{bmatrix}. \qquad (4.5.35)$$

We consider now a weighted vector norm $\|w\|_{\tilde{\mathcal{E}}_1} = \sqrt{w^T\tilde{\mathcal{E}}_1 w}$ for $w \in \mathbb{R}^{\eta+m}$. It is well defined since $\tilde{\mathcal{E}}_1$ is symmetric and positive definite. This norm is equivalent to the Euclidean norm $\|w\|$ due to the inequalities

$$\sqrt{\lambda_{\min}(\tilde{\mathcal{E}}_1)}\|w\| \leqslant \|w\|_{\tilde{\mathcal{E}}_1} \leqslant \sqrt{\lambda_{\max}(\tilde{\mathcal{E}}_1)}\|w\|, \qquad (4.5.36)$$

where $\lambda_{\min}(\tilde{\mathcal{E}}_1)$ and $\lambda_{\max}(\tilde{\mathcal{E}}_1)$ denote the smallest and largest eigenvalues of $\tilde{\mathcal{E}}_1$, respectively. Using (4.5.35) and the definition of $L_2[\tilde{f}]$, we can estimate

$$\begin{aligned}
\|\varepsilon_1(t)\|_{\tilde{\mathcal{E}}_1}\frac{d}{dt}\|\varepsilon_1(t)\|_{\tilde{\mathcal{E}}_1} &= \langle\varepsilon_1(t),\dot{\varepsilon}_1(t)\rangle_{\tilde{\mathcal{E}}_1} = \langle\varepsilon_1(t),\tilde{\mathcal{E}}_1\dot{\varepsilon}_1(t)\rangle \\
&= \left\langle\varepsilon_1(t), \tilde{\mathcal{A}}_{1l}\varepsilon_1(t) + \begin{bmatrix} U_{a_1}^T f_1(U_{a_1}\tilde{a}_1(t)) - \hat{f}_1(\hat{a}_1(t)) \\ 0 \end{bmatrix}\right\rangle \\
&= \left\langle\varepsilon_1(t), \tilde{\mathcal{A}}_{1l}\varepsilon_1(t) + \begin{bmatrix} U_{a_1}^T f_1(U_{a_1}\tilde{a}_1(t)) - U_{a_1}^T f_1(U_{a_1}\hat{a}_1(t)) \\ 0 \end{bmatrix}\right\rangle \\
&\quad + \left\langle\varepsilon_1(t), \begin{bmatrix} U_{a_1}^T f_1(U_{a_1}\hat{a}_1(t)) - \hat{f}_1(\hat{a}_1(t)) \\ 0 \end{bmatrix}\right\rangle \\
&= \langle\tilde{x}_1(t)-\hat{x}_1(t), \tilde{\mathcal{A}}_1(\tilde{x}_1(t))\tilde{x}_1(t) - \tilde{\mathcal{A}}_1(\hat{x}_1(t))\hat{x}_1(t)\rangle \\
&\quad + \left\langle\tilde{a}_1(t)-\hat{a}_1(t), U_{a_1}^T f_1(U_{a_1}\hat{a}_1(t)) - \hat{f}_1(\hat{a}_1(t))\right\rangle \\
&\leqslant L_2[\tilde{f}]\|\varepsilon_1(t)\|^2 + \beta\|\varepsilon_1(t)\|.
\end{aligned}$$

In the last inequality, we used the estimate for the DEIM error as presented in Theorem 2.24 and define

$$\beta = \|(\mathcal{S}_{\mathcal{K}}^T U_f)^{-1}\|_2 \sum_{i=\kappa+1}^{n_s} \sigma_i(\mathcal{X}_f).$$

Furthermore, taking into account (4.5.36), we obtain

$$\begin{aligned}\frac{d}{dt}\|\varepsilon_1(t)\|_{\tilde{\mathcal{E}}_1} &\leqslant L_2[\tilde{f}]\frac{\|\varepsilon_1(t)\|^2}{\|\varepsilon_1(t)\|_{\tilde{\mathcal{E}}_1}} + \beta\frac{\|\varepsilon_1(t)\|}{\|\varepsilon_1(t)\|_{\tilde{\mathcal{E}}_1}}\\ &\leqslant \frac{L_2[\tilde{f}]}{\lambda_{\min}(\tilde{\mathcal{E}}_1)}\|\varepsilon_1(t)\|_{\tilde{\mathcal{E}}_1} + \frac{\beta}{\sqrt{\lambda_{\min}(\tilde{\mathcal{E}}_1)}}.\end{aligned} \tag{4.5.38}$$

Using the comparison lemma [WSH14], we have

$$\begin{aligned}\|\varepsilon_1(t)\|_{\tilde{\mathcal{E}}_1} &\leqslant \int_0^t \frac{\beta}{\sqrt{\lambda_{\min}(\tilde{\mathcal{E}}_1)}} e^{\int_s^t \frac{L_2[\tilde{f}]}{\lambda_{\min}(\tilde{\mathcal{E}}_1)}\,d\tau}\,ds\\ &= \frac{\beta}{\sqrt{\lambda_{\min}(\tilde{\mathcal{E}}_1)}}\int_0^t e^{\frac{L_2[\tilde{f}]}{\lambda_{\min}(\tilde{\mathcal{E}}_1)}(t-s)}\,ds\\ &= \frac{\beta\sqrt{\lambda_{\min}(\tilde{\mathcal{E}}_1)}}{L_2[\tilde{f}]}\Big(e^{\frac{L_2[\tilde{f}]}{\lambda_{\min}(\tilde{\mathcal{E}}_1)}t} - 1\Big).\end{aligned}$$

Finally, we use the norm equivalence (4.5.36) and obtain

$$\|\varepsilon_1(t)\| \leqslant \frac{1}{\sqrt{\lambda_{\min}(\tilde{\mathcal{E}}_1)}}\|\varepsilon_1(t)\|_{\tilde{\mathcal{E}}_1} \leqslant \frac{\beta}{L_2[\tilde{f}]}\Big(e^{\frac{L_2[\tilde{f}]}{\lambda_{\min}(\tilde{\mathcal{E}}_1)}t} - 1\Big).$$ □

Next, we present computable estimates on the logarithmic Lipschitz constant $L_2[\tilde{f}]$.

Theorem 4.19. *The logarithmic Lipschitz constant* $L_2[\tilde{f}]$ *can be estimated as*

1. $L_2[\tilde{f}] \leqslant -m_\nu \lambda_{\max}(\tilde{R}^T K_1 \tilde{R}) =: \mu_1$ *with* $\tilde{R}$ *given in* (4.5.30) *and* K_1 *in* (4.3.10),
2. $L_2[\tilde{f}] \leqslant \lambda_{\max}(\tilde{\mathcal{A}}_{1l}) - m_{\nu,1}\lambda_{\max}(U_{a_1}^T K_{1,n} U_{a_1}) =: \mu_2$ *where the entries of* $K_{1,n}$ *are given by* $(K_{1,n})_{i,j} = \int_{\Omega_1} \nabla\psi_i \cdot \nabla\psi_j \, d\xi$, $i,j = 1,\dots,n_1$.

Proof. Similarly to the proof of Theorem 4.3, we can show for the function

$$\tilde{g}(\tilde{x}_1) = \tilde{R}^T \mathcal{K}(\tilde{R}\tilde{x}_1)\tilde{R}\tilde{x}_1$$

that

$$\begin{aligned}\langle \tilde{g}(\tilde{x}_1) - \tilde{g}(\tilde{w}), \tilde{x}_1 - \tilde{w}\rangle &= \langle \mathcal{K}(\tilde{R}\tilde{x}_1)\tilde{R}\tilde{x}_1 - \mathcal{K}(\tilde{R}\tilde{w})\tilde{R}\tilde{w}, \tilde{R}(\tilde{x}_1 - \tilde{w})\rangle\\ &\geqslant m_\nu(\tilde{x}_1 - \tilde{w})^T \tilde{R}^T K_1 \tilde{R}(\tilde{x}_1 - \tilde{w})\end{aligned}$$

for all $\tilde{x}_1, \tilde{w} \in \mathbb{R}^{\eta+m}$. Then using the relation $\tilde{f}(\tilde{x}_1) = \tilde{\mathcal{A}}_1(\tilde{x}_1)\tilde{x}_1 = -\tilde{R}^T \mathcal{K}(\tilde{R}\tilde{x}_1)\tilde{R}\tilde{x}_1$, we obtain

$$\begin{aligned} L_2[\tilde{f}] &= \sup_{\substack{\tilde{x}_1, \tilde{w} \in \mathbb{R}^{\eta+m} \\ \tilde{x}_1 \neq \tilde{w}}} \frac{\langle \tilde{x}_1 - \tilde{w}, \tilde{f}(\tilde{x}_1) - \tilde{f}(\tilde{w})\rangle}{\|\tilde{x}_1 - \tilde{w}\|^2} \\ &\leqslant -m_{\boldsymbol{\nu}} \sup_{\substack{\tilde{x}_1, \tilde{w} \in \mathbb{R}^{\eta+m} \\ \tilde{x}_1 \neq \tilde{w}}} \frac{\langle \tilde{R}(\tilde{x}_1 - \tilde{w}), K_1 \tilde{R}(\tilde{x}_1 - \tilde{w})\rangle}{\|\tilde{x}_1 - \tilde{w}\|^2} \\ &= -m_{\boldsymbol{\nu}} \lambda_{\max}(\tilde{R}^T K_1 \tilde{R}). \end{aligned}$$

2. We first split $\tilde{f}(\tilde{x}_1) = \tilde{\mathcal{A}}_{1l}\tilde{x}_1 + \begin{bmatrix} U_{a_1}^T f_1(U_{a_1}\tilde{a}_1) \\ 0 \end{bmatrix}$. Then using the definition of the logarithmic Lipschitz constant for $\tilde{f}$, we get

$$\begin{aligned} L_2[\tilde{f}] &= \sup_{\substack{\tilde{x}_1, \tilde{w} \in \mathbb{R}^{\eta+m} \\ \tilde{x}_1 \neq \tilde{w}}} \frac{\langle \tilde{x}_1 - \tilde{w}, \tilde{\mathcal{A}}_{1l}(\tilde{x}_1 - \tilde{w}) + \begin{bmatrix} U_{a_1}^T (f_1(U_{a_1}\tilde{a}_1) - f_1(U_{a_1}\tilde{w}_1)) \\ 0 \end{bmatrix}\rangle}{\|\tilde{x}_1 - \tilde{w}\|^2} \\ &\leqslant L_2\left[\tilde{\mathcal{A}}_{1l}\right] + \sup_{\substack{\tilde{x}_1, \tilde{w} \in \mathbb{R}^{\eta+m} \\ \tilde{x}_1 \neq \tilde{w}}} \frac{\langle \tilde{x}_1 - \tilde{w}, \begin{bmatrix} U_{a_1}^T (f_1(U_{a_1}\tilde{a}_1) - f_1(U_{a_1}\tilde{w}_1)) \\ 0 \end{bmatrix}\rangle}{\|\tilde{x}_1 - \tilde{w}\|^2}. \end{aligned}$$

Since $\tilde{\mathcal{A}}_{1l}$ is symmetric, it holds for the logarithmic Lipschitz constant $L_2[\tilde{\mathcal{A}}_{1l}] = \lambda_{\max}(\tilde{\mathcal{A}}_{1l})$. Furthermore, it follows from $g_1(U_{a_1} a_1) = \mathcal{K}_{11,n}(U_{a_1} a_1)U_{a_1} a_1$ with $\mathcal{K}_{11,n}$ as in (4.3.13) and

$$(\boldsymbol{\nu}_1(\|\varphi\|)\varphi - \boldsymbol{\nu}_1(\|\tilde{\varphi}\|)\tilde{\varphi}) \cdot (\varphi - \tilde{\varphi}) \geqslant m_{\boldsymbol{\nu},1}\|\varphi - \tilde{\varphi}\|^2,$$

see [Pec04, proof of Lemma 2.8] that

$$\begin{aligned} &\langle \begin{bmatrix} U_{a_1} \\ 0 \end{bmatrix} (\tilde{x}_1 - \tilde{w}), \begin{bmatrix} g_1(U_{a_1}\tilde{a}_1) - g_1(U_{a_1}\tilde{w}_1) \\ 0 \end{bmatrix}\rangle \\ &= \int_{\Omega_1} \left(\boldsymbol{\nu}_1(\|\nabla \sum_{i=1}^{\eta} \tilde{\alpha}_i \tilde{\psi}_i(\xi)\|)\nabla \sum_{i=1}^{\eta} \tilde{\alpha}_i \tilde{\psi}_i(\xi) - \boldsymbol{\nu}_1(\|\nabla \sum_{i=1}^{\eta} \tilde{\omega}_i \tilde{\psi}_i(\xi)\|)\nabla \sum_{i=1}^{\eta} \tilde{\omega}_i \tilde{\psi}_i(\xi) \right) \\ &\cdot \nabla \sum_{i=1}^{\eta} (\tilde{\alpha}_i - \tilde{\omega}_i)\tilde{\psi}_i(\xi)\, d\xi \\ &\geqslant m_{\boldsymbol{\nu},1} \int_{\Omega_1} \nabla \sum_{i=1}^{\eta} (\tilde{\alpha}_i - \tilde{\omega}_i)\tilde{\psi}_i(\xi) \cdot \nabla \sum_{i=1}^{\eta} (\tilde{\alpha}_i - \tilde{\omega}_i)\tilde{\psi}_i(\xi)\, d\xi \\ &= m_{\boldsymbol{\nu},1}(\tilde{a}_1 - \tilde{w}_1)^T U_{a_1}^T K_{1,n} U_{a_1}(\tilde{a}_1 - \tilde{w}_1) \end{aligned}$$

with $\tilde{w}_1 = [\tilde{\omega}_1, \dots, \tilde{\omega}_\eta]^T$ and $[\tilde{\psi}_1, \dots, \tilde{\psi}_\eta] = [\psi_1, \dots, \psi_{n_1}]U_{a_1}$. Using the relation $f_1(a_1) = -\mathcal{K}_{11,n}(a_1)a_1$, this results in

$$\begin{aligned} L_2[\tilde{f}] &\leqslant \lambda_{\max}(\tilde{\mathcal{A}}_{1l}) + \sup_{\substack{\tilde{a}_1, \tilde{w}_1 \in \mathbb{R}^{\eta+m} \\ \tilde{a}_1 \neq \tilde{w}_1}} \frac{\langle \tilde{a}_1 - \tilde{w}_1, U_{a_1}^T (f_1(U_{a_1}\tilde{a}_1) - f_1(U_{a_1}\tilde{w}_1))\rangle}{\|\tilde{a}_1 - \tilde{w}_1\|^2} \\ &\leqslant \lambda_{\max}(\tilde{\mathcal{A}}_{1l}) - m_{\nu,1} \sup_{\substack{\tilde{a}_1, \tilde{w}_1 \in \mathbb{R}^{\eta+m} \\ \tilde{a}_1 \neq \tilde{w}_1}} \frac{\langle U_{a_1}(\tilde{a}_1 - \tilde{w}_1), K_{1,n}U_{a_1}(\tilde{a}_1 - \tilde{w}_1)\rangle}{\|\tilde{a}_1 - \tilde{w}_1\|^2} \\ &= \lambda_{\max}(\tilde{\mathcal{A}}_{1l}) - m_{\nu,1}\lambda_{\max}(U_{a_1}^T K_{1,n} U_{a_1}). \end{aligned}$$

□

It follows from Theorem 4.19 that

$$L_2[f] \leqslant \min(\mu_1, \mu_2) := \mu.$$

Using this bound we obtain from (4.5.38) that

$$\frac{d}{dt}\|\varepsilon_1(t)\|_{\tilde{\mathcal{E}}_1} \leqslant \frac{\mu}{\lambda_{\min}(\tilde{\mathcal{E}}_1)}\|\varepsilon_1(t)\| + \frac{\beta}{\sqrt{\lambda_{\min}(\tilde{\mathcal{E}}_1)}}.$$

Similarly to the proof of Theorem 4.18, we estimate

$$\|\varepsilon_1(t)\| \leqslant \frac{\beta}{\mu}\left(e^{\frac{\mu t}{\lambda_{\min}(\tilde{\mathcal{E}}_1)}} - 1\right) =: \epsilon(t). \tag{4.5.39}$$

4.6 Numerical example

In this section, we present some results of numerical experiments for model order reduction of linear and nonlinear MQS for a single-phase 2D transformer. For the mesh generation and the FEM discretization, we used the software package FEniCS[1] of version 1.4. The time integration of the full models is done by the sparse DAE solver PyDaeSI provided by Caren Tischendorf, whereas the reduced-order dense systems are solved by the implicit differential-algebraic (IDA) solver from the simulation package Assimulo[2]. Both solvers are based on the backward differentiation formula methods. We use them according to the dense or sparse structure of the problem. The computations were performed on a computer with an Intel(R) Core(TM) i7-3720QM processor with 2.60GHz.

[1]http://fenicsproject.org
[2]http://www.jmodelica.org/assimulo

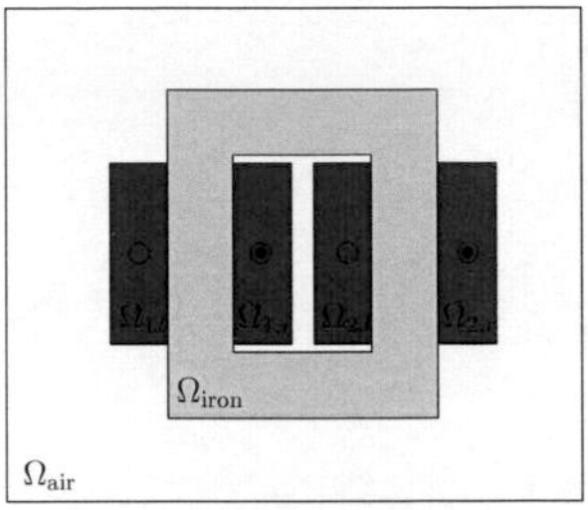

(a) Geometry

Subdomains
$\Omega_1 = \Omega_{\text{iron}},\ \ \Omega_2 = \Omega_{1,l} \cup \Omega_{1,r} \cup \Omega_{2,l} \cup \Omega_{2,r} \cup \Omega_{\text{air}}$
Dimensions
$n_a = 51543,\ n_1 = 19688,\ n_2 = 31855,\ m = 2$
Model parameters
$\boldsymbol{\sigma}_1 = 5 \cdot 10^5\, \Omega^{-1}\text{m}^{-1}$
$\boldsymbol{\nu}_2 = 1\, \text{AmV}^{-1}\text{s}^{-1}$
$N_1 = 358,\ N_2 = 206$
$\mathcal{S}_1 = \mathcal{S}_2 = 1.12 \cdot 10^{-2}\ \text{m}^2$
$\mathcal{R} = \text{diag}(\rho_1, \rho_2),\ \rho_j = 1.73 \cdot 10^{-8} \frac{N_j}{\mathcal{S}_j},\ j = 1,2$

(b) Dimensions and parameters

Figure 4.4: Single-phase 2D transformer

4.6.1 2D transformer model

We consider a single-phase 2D transformer with an iron core and two coils of wire in an air domain as shown in Figure 4.4(a), see [Sch11] for detailed description and geometry data. The winding function has two components given by

$$\boldsymbol{\chi}_{\text{str},3,j}(\xi) = \begin{cases} -\frac{N_j}{\mathcal{S}_j} & \text{for } \xi \in \Omega_{j,l}, \\ \frac{N_j}{\mathcal{S}_j} & \text{for } \xi \in \Omega_{j,r}, \\ 0 & \text{for } \xi \in \Omega \setminus (\Omega_{j,l} \cup \Omega_{j,r}), \end{cases}$$

where N_j are the number of coil turns and $\mathcal{S}_j$ is the cross section area for $j = 1, 2$. We insert a thin layer of air between the coils and the iron core to avoid unphysical effects caused by the FEM discretization. Using the linear Lagrange elements on a uniform triangular mesh, we obtain a semidiscretized MQS system (4.3.5) with $\mathcal{X}_1 = 0$. It should, however, be noted that our model reduction method works also without the air gap. The model parameters and dimensions are presented in Figure 4.4(b).

4.6.2 Balanced truncation

First, we apply BT to the linear transformer model obtained for the constant magnetic reluctivity $\boldsymbol{\nu}_1 = 14872\, \text{AmV}^{-1}\text{s}^{-1}$. The controllability Gramian G_c is approximated by a low-rank matrix $G_c \approx \tilde{Z}_c \tilde{Z}_c^T$ with $\tilde{Z}_c \in \mathbb{R}^{(n_1+m)\times k}$, where $n_1 + m = 19670$ and $k = 56$. The Hankel singular values are presented in Figure 4.5. One can see that they decay very rapidly implying that the system can be well approximated by a reduced model of small dimension. We choose the reduced dimension $\eta = 3$ and get an error bound

$$\begin{aligned} \|\mathbf{G} - \tilde{\mathbf{G}}\|_{\mathcal{H}_\infty} &\leqslant 2(\varpi_{\eta+1} + \ldots + \varpi_{n_1+m}) \\ &\leqslant 2(\varpi_{\eta+1} + \ldots + \varpi_{k-1} + (n_1 + m - k + 1)\varpi_k) =: \gamma \end{aligned}$$

with

$$\gamma = 2.3993 \cdot 10^{-4}. \tag{4.6.1}$$

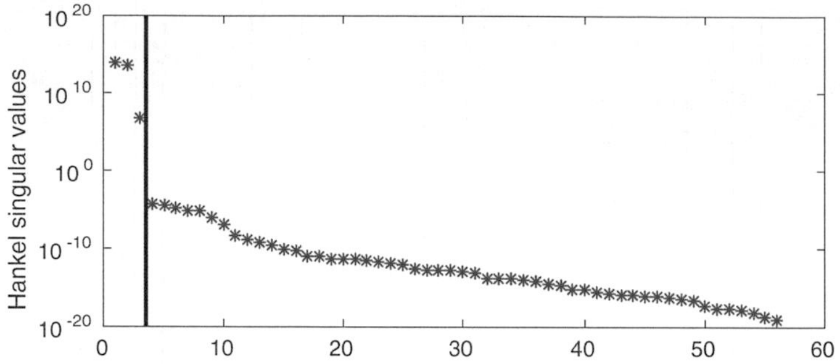

Figure 4.5: Hankel singular values

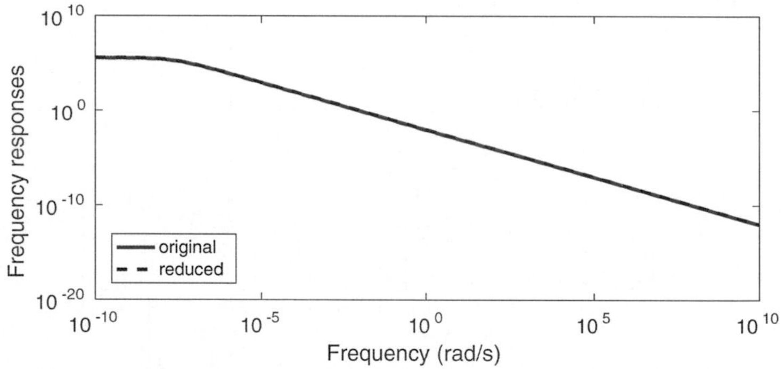

Figure 4.6: Linear MQS model: frequency responses of the original and the reduced systems

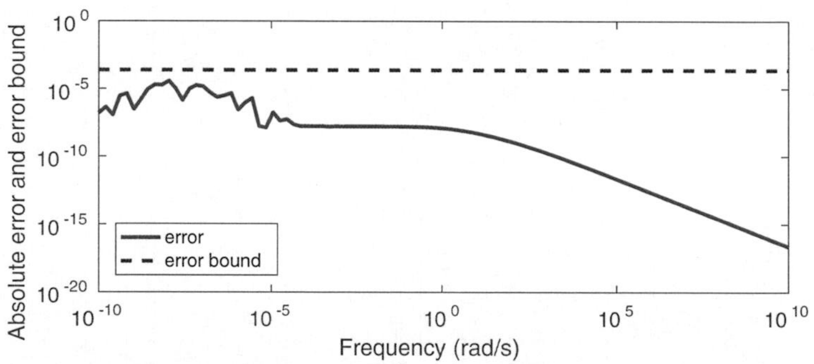

Figure 4.7: Linear MQS model: absolute error in the frequency domain and error bound γ

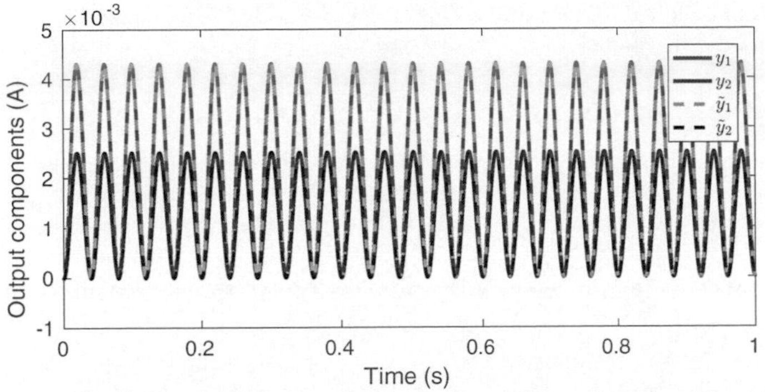

Figure 4.8: Linear MQS model: output components of the original and the reduced systems

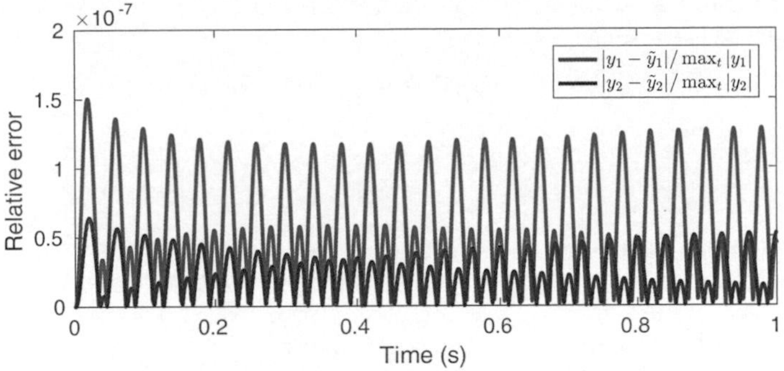

Figure 4.9: Linear MQS model: relative error in the output components

Figure 4.6 shows the spectral norms of the frequency responses $\|\mathbf{G}(i\omega)\|_2$ and $\|\tilde{\mathbf{G}}(i\omega)\|_2$ of the full and the reduced models for the frequency range $\omega \in [10^{-10}, 10^{10}]$. In Figure 4.7, the error bound γ and the absolute error $\|\mathbf{G}(i\omega) - \tilde{\mathbf{G}}(i\omega)\|_2$ are presented. For the input

$$u(t) = \begin{bmatrix} 80 \cdot \sin(50\pi t) \\ 20 \cdot \sin(50\pi t) \end{bmatrix}$$

and a zero initial condition, we have computed the outputs $y(t) = [y_1(t), y_2(t)]^T$ and $\tilde{y}(t) = [\tilde{y}_1(t), \tilde{y}_2(t)]^T$ of the full and reduced systems, respectively, on the time interval $[0, 1]s$. Figure 4.8 shows the components of these outputs. The relative errors $|y_j(t) - \tilde{y}_j(t)|/\max_{\tau\in[0,1]} |y_j(\tau)|$, $j = 1, 2$, in the output components are presented in Figure 4.9.

One can see that the errors in time and frequency domains are small, while the reduced system can be computed 1500 times faster than the original system. Taking into account the reduction time, we get a speedup of 46 for a single simulation of the reduced-order model.

4.6.3 POD and DEIM

Next, we examine model reduction of the 2D nonlinear transformer model as in Section 4.6.1, where the magnetic reluctivity of the iron core is given by

$$\nu_1(\varrho) = 3.8 \exp(2.14\varrho^2) + 396.2\,\mathrm{AmV^{-1}s^{-1}}.$$

The snapshots were collected with the training input $u(t)$ and the reduced models were tested with the input $u_{test}(t)$ given by

$$u(t) = \begin{bmatrix} 45.5 \cdot 10^3 \sin(900\pi t) \\ 77 \cdot 10^3 \sin(1700\pi t) \end{bmatrix}, \quad u_{test}(t) = \begin{bmatrix} 46.5 \cdot 10^3 \sin(1010\pi t) \\ 78 \cdot 10^3 \sin(1900\pi t) \end{bmatrix}. \tag{4.6.2}$$

Figures 4.10 and 4.11 shows the singular values σ_j^{POD} and σ_j^{DEIM} of the snapshot matrices $\mathcal{X}_{a_1}$ and $\mathcal{X}_f$ of POD and DEIM, respectively, and the eigenvalues λ_j^{MDEIM} of the snapshot matrix $\mathcal{X}_J$ of MDEIM are presented in Figure 4.12. The reduced dimensions were chosen as

$$\text{POD: } \eta = 35, \qquad \text{DEIM: } \kappa = 9, \qquad \text{MDEIM: } \rho = 3,$$

with the relations

$$\frac{\sigma_{36}^{POD}}{\sigma_1^{POD}} = 1.14 \cdot 10^{-7}, \qquad \frac{\sigma_{10}^{DEIM}}{\sigma_1^{DEIM}} = 1.40 \cdot 10^{-3}, \qquad \frac{\lambda_4^{MDEIM}}{\lambda_1^{MDEIM}} = 1.07 \cdot 10^{-5}.$$

The reduced model has dimension $r = 37$. The Jacobi matrix $J_{f_1} \in \mathbb{R}^{n_1 \times n_1}$ has 97464 nonzero entries, whereas $\mathcal{S}_\mathcal{K}^T J_{f_1} \in \mathbb{R}^{9 \times n_1}$ has 58 nonzero entries. Using MDEIM the evaluation of only 3 entries of $\mathcal{S}_\mathcal{K}^T J_{f_1}$ are used. The extended index sets $\mathcal{K}_{ext}$ and $\mathcal{J}_{ext}$ have $|\mathcal{K}_{ext}| = 47$ and $|\mathcal{J}_{ext}| = 18$ entries, respectively.

Figure 4.13 shows the output components of the original and POD-DEIM-MDEIM reduced systems. The relative errors

$$\Delta(y, \tilde{y}) = \sqrt{\sum_{i=1}^{2} \left(\frac{y_i(t) - \tilde{y}_i(t)}{\max\limits_{\tau \in [0, 0.01]} |y_i(\tau)|} \right)^2} \tag{4.6.3}$$

of the POD, POD-DEIM and POD-DEIM-MDEIM reduced systems can be seen in Figure 4.14. Note that for the POD-DEIM variant, we use the affine representation of the Jacobi matrix $J_{\hat{f}_1}$ given in (4.5.24). The errors introduced by DEIM and MDEIM can be neglected in comparison to the POD error. In Table 4.1, we present the computation time required to construct the snapshot matrices $\mathcal{X}_{a_1}$, $\mathcal{X}_f$ and $\mathcal{X}_J$, the reduction time which includes the computation of the projection matrices and the time-independent matrices as in Step 6 of Figure 4.3, as well as the simulation time for the reduced-order models determined by the POD, POD-DEIM and POD-DEIM-MDEIM methods. Comparing the simulation time 3269s for the original system to that for the reduced-order models, we achieve a speed up of about 20 for the POD model, 277 for the POD-DEIM model and 314 for the POD-DEIM-MDEIM model.

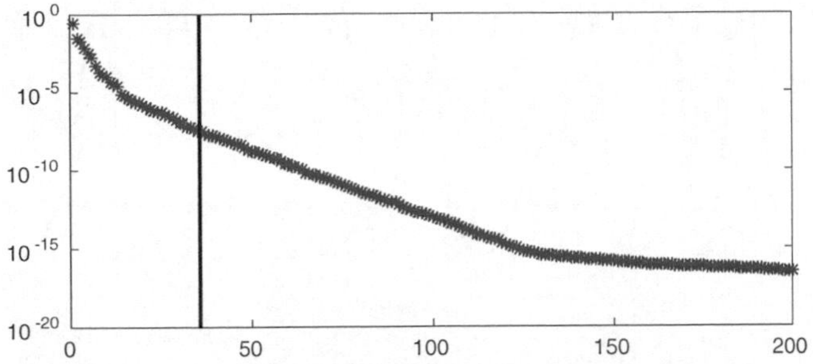

Figure 4.10: Nonlinear MQS model: POD singular values

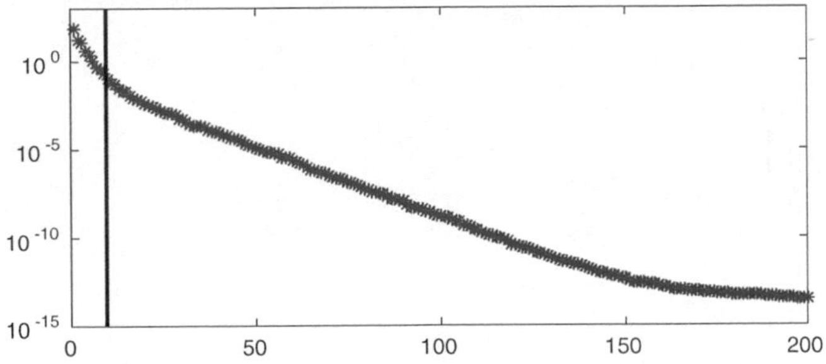

Figure 4.11: Nonlinear MQS model: DEIM singular values

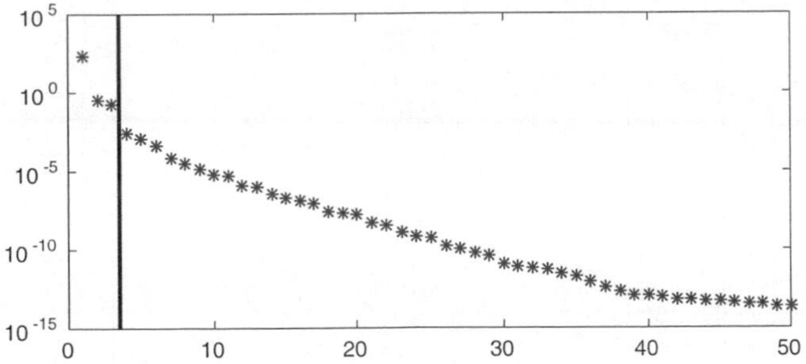

Figure 4.12: Nonlinear MQS model: MDEIM eigenvalues

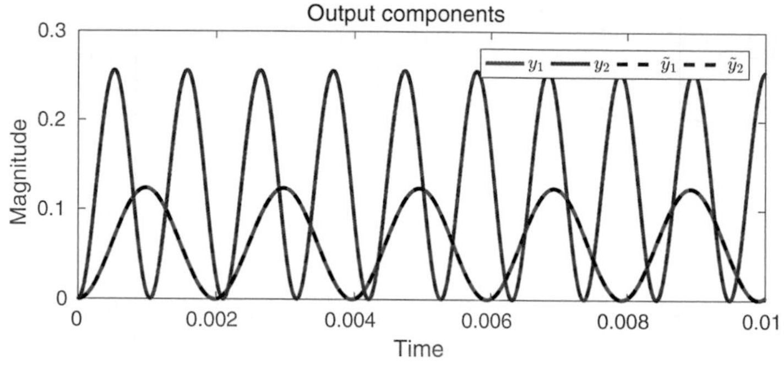

Figure 4.13: Nonlinear MQS model: output components of the original and reduced models

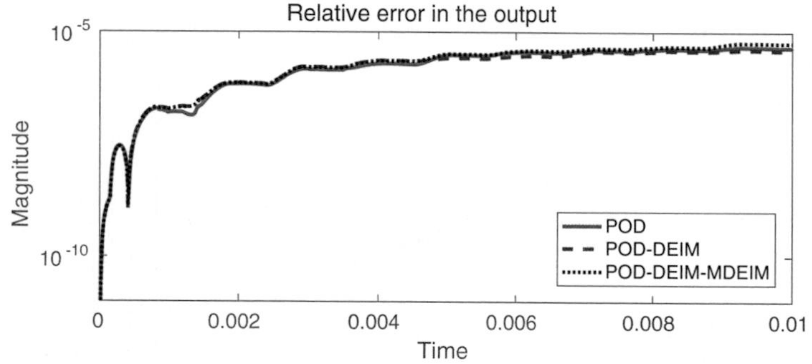

Figure 4.14: Nonlinear MQS model: relative errors in the output

	POD	POD-DEIM	POD-DEIM-MDEIM
Computation time for the snapshots			
$\mathcal{X}_{a_1}$	2870.7	2870.7	2870.7
$\mathcal{X}_f$		37.5	37.5
$\mathcal{X}_J$			366.5
Reduction time (projection matrices U_{a_1} and U_f, basis matrices $V_1, \dots, V_p$, time-independent matrices)	54.8	197.4	357.1
Simulation time for the reduced-oder model	162.8	11.8	10.4

Table 4.1: Computation time (in seconds) for the POD, POD-DEIM and POD-DEIM-MDEIM methods

4.6.4 Passivity enforcing

We now verify the DEIM error bound ϵ from Section 4.5.5 required for the passivity enforcement of the POD-DEIM reduced system. In Table 4.2, we collect the constants involved in (4.5.34) and (4.5.39). Figure 4.15 shows the absolute error $\|\varepsilon_1(t)\| = \|\tilde{x}_1(t) - \hat{x}_1(t)\|$ and the error bounds $\epsilon_i(t)$ as in (4.5.39) computed with $\mu = \mu_i$ for $i = 1, 2$. One can see that the error bound $\epsilon_1(t)$ overestimates the true error by about two orders of magnitude, while the error bound $\epsilon_2(t)$ is quite sharp. In Figure 4.16, we present the output error $\|\tilde{y}(t) - \hat{y}(t)\|$ and the error bounds $\epsilon_i^y(t) = \|\tilde{\mathcal{C}}_1\|_2 \epsilon_i(t)$ for $i = 1, 2$.

To enforce passivity, the output of the POD-DEIM reduced model (4.5.20) with $u = u_{test}$ as in (4.6.2) is perturbed by δu_{test}, where δ is given in (4.5.33). Figure 4.17 shows the components of the perturbation $\delta(t)u_{test}(t)$. Finally, Figure 4.18 shows the relative errors $\Delta(y, \hat{y})$ and $\Delta(y, y_\delta)$ defined in (4.6.3). We see that the error of the perturbed POD-DEIM system (4.5.31) is only slightly larger than that of the POD-DEIM system (4.5.20).

$\beta = 4.1467$
$\lambda_{\min}(\tilde{\mathcal{E}}_1) = 0.1969$
$\mu_1 = -7.7916, \quad m_\nu = \min(396, 1) = 1, \ \lambda_{\max}(\tilde{R}^T K_1 \tilde{R}) = 7.7916$
$\mu_2 = -2755, \quad m_{\nu,1} = 396, \ \lambda_{\max}(\tilde{\mathcal{A}}_{1l}) = -1.4307 \cdot 10^{-4}, \ \lambda_{\max}(U_{a_1}^T K_{1,n} U_{a_1}) = 6.9569$
$\|\tilde{\mathcal{C}}_1\|_2 = 1.591 \cdot 10^{-4}$

Table 4.2: Constants for the error bound

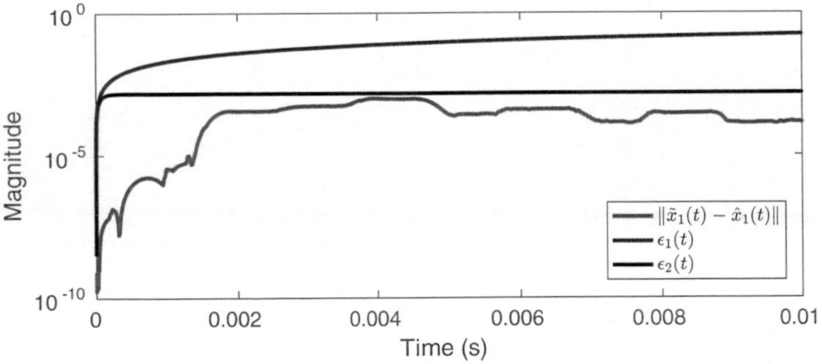

Figure 4.15: Absolute error in the state and the error estimators

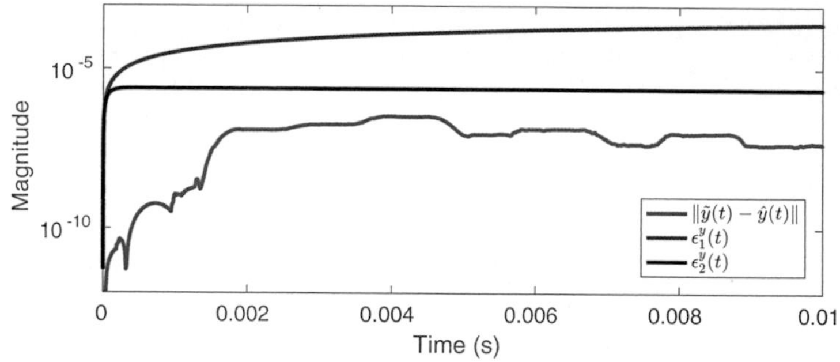

Figure 4.16: Absolute error in the output and the error estimators

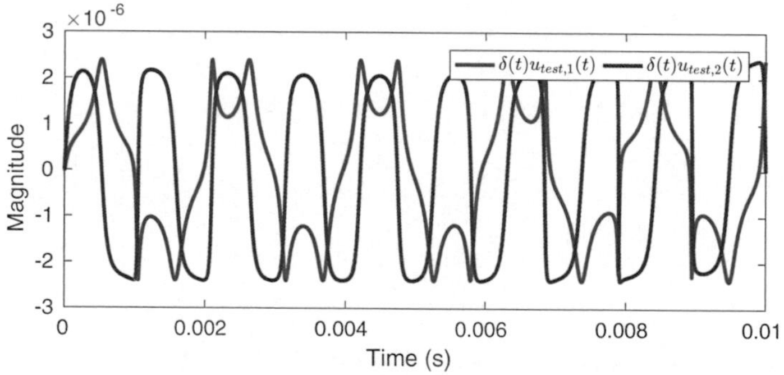

Figure 4.17: Perturbation $\delta(t)u_{test}(t)$

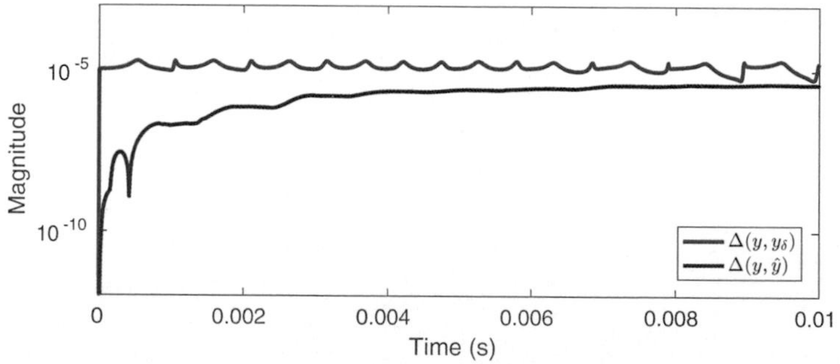

Figure 4.18: Relative errors in the output

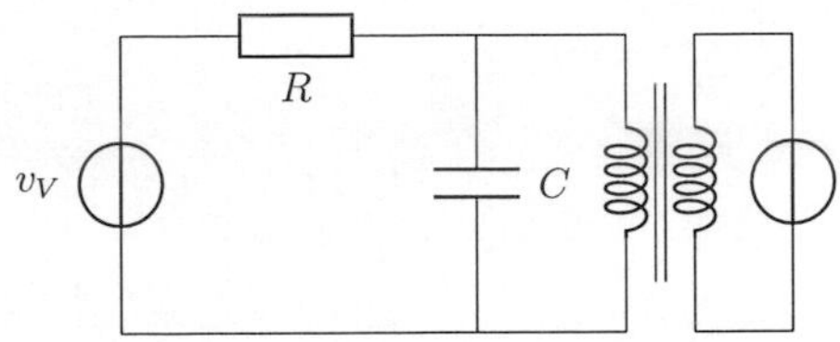

Figure 4.19: Coupled MQS-circuit system

4.6.5 Dynamic iteration for the coupled MQS-circuit system

Finally, we consider a transformer coupled to a simple electrical circuit as shown in Figure 4.19 and compute the solutions of the resulting coupled MQS-circuit system using a monolithic approach and dynamic iteration presented in Section 3.1. Furthermore, we approximate the linear MQS subsystem by a reduced-order model determined by BT and repeat the simulation using again the monolithic approach and dynamic iteration. Note that a similar coupled MQS-circuit system has been solved using a multirate time integration scheme in [HKBB+18].

First, we briefly introduce circuit equations in the modified nodal analysis (MNA) formulation, see [HRB75, Rei14, Ria08, VS94] for further information. An electrical circuit is modelled as a directed graph whose vertices correspond to the nodes and whose branches correspond to the circuit elements like resistors, inductors and capacitors. Let n_n+1 and n_b be the numbers of nodes and branches, respectively. Then the network topology is described by an incidence matrix $\mathcal{A}_0 \in \mathbb{R}^{n_n \times n_b}$ with entries

$$(\mathcal{A}_0)_{i,j} = \begin{cases} 1, & \text{if branch } j \text{ leaves node } i, \\ -1, & \text{if branch } j \text{ enters node } i, \\ 0, & \text{otherwise.} \end{cases}$$

By removing a row from $\mathcal{A}_0$ corresponding to a ground node and reordering the branches accordingly to the type of circuit components, one gets a reduced incidence matrix

$$\mathcal{A} = \begin{bmatrix} A_R & A_L & A_C & A_V & A_I & A_M \end{bmatrix}$$

of full rank, where the subscripts R, L, C, V, I and M stand for resistors, inductors, capacitors, voltage sources, current sources and electromagnetic devices, respectively. Using Kirchhoff's current and voltage laws and constitutive relations characterizing the circuit components, the dynamics of the lumped element circuit can be described based on MNA by a DAE system

$$\begin{aligned} A_C \tfrac{d}{dt} q_C(A_C^T\rho) + A_R g_R(A_R^T\rho) + A_L i_L + A_V i_V + A_I i_I + A_M i_M &= 0, \\ \tfrac{d}{dt}\varphi_L(i_L) - A_L^T\rho &= 0, \\ A_V^T\rho - v_V &= 0, \end{aligned}$$

where ρ is the vector of node potentials, i_L, i_V, i_I and i_M are the vectors of currents through inductors, voltage sources, current sources and electromagnetic devices, respectively, and v_V is the voltage vector for voltage sources. The functions g_R,

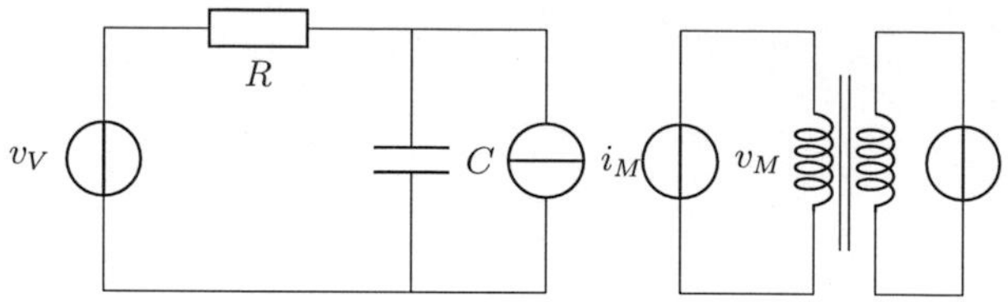

Figure 4.20: Decoupled System

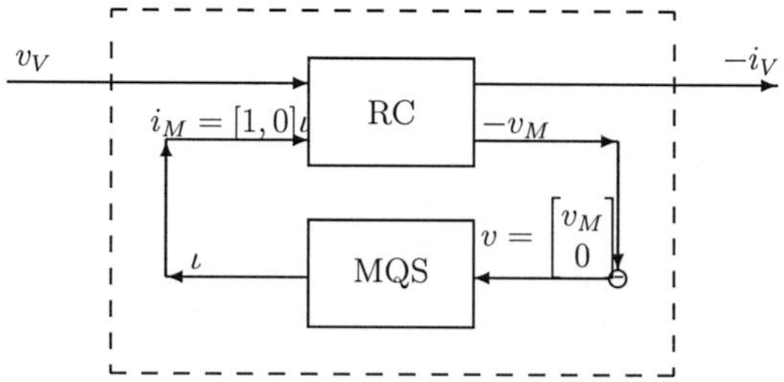

Figure 4.21: The coupled RC-MQS system

φ_L and q_C describe voltage-current characteristics of resistors, magnetic fluxes in inductors and capacitor charges, respectively.

We consider now the coupled MQS-circuit in Figure 4.19. In order to decouple this system into two subsystems, we insert controlled current and voltage sources with current i_M and voltage v_M, respectively. The resulted decoupled system is presented in Figure 4.20. The MNA equation for the RC circuit is given by

$$\begin{aligned} A_C C A_C^T \rho + A_R R^{-1} A_R^T \rho + A_V i_V + A_M i_M &= 0, \\ A_V^T \rho - v_V &= 0, \end{aligned}$$

with

$$A_C = \begin{bmatrix} 0 \\ 1 \end{bmatrix}, \quad A_R = \begin{bmatrix} -1 \\ 1 \end{bmatrix}, \quad A_V = \begin{bmatrix} 1 \\ 0 \end{bmatrix}, \quad A_M = \begin{bmatrix} 0 \\ 1 \end{bmatrix},$$

the resistance $R = 10^3$, the capacitance $C = 1$, and the voltage $v_V(t) = 50\sin(50t)$ at the voltage source. For the transformer, we use the MQS model (4.2.10) with piecewise constant magnetic reluctivity. The geometry and material parameters are as in Section 4.6.1. The block schema for the coupled MQS-circuit system is given in Figure 4.21. The internal inputs of the MQS system and the circuit system are given by $v = [v_M, 0]^T$ and $i_M = [1,0]\iota$, respectively, whereas the internal outputs have the form ι and $-v_M$, respectively. The external input and output of the coupled system are given by v_V and $-i_V$, respectively. The semidiscretized MQS system was approximated by a reduced-order model as described in Section 4.6.2. Note that the error bound γ in (4.6.1) for the reduced MQS subsystem can be used to estimate the error in the coupled system, see [RS07] for details.

In Table 4.3, we present the computation time for four different simulation approaches: monolithic, dynamic iteration, monolithic combined with model reduction and dynamic iteration combined with model reduction. We include there also the time for computing the reduced-order MQS model. We used the Jacobi-type dynamic iteration method with $n_T = 10$ macro time steps and $K_m = 2$ iterations on each time window. One can see that the dynamic iteration without model reduction is about twice faster than the monolithic approach. Using model reduction reduces the simulation time significantly even if we take into account the reduction time. Comparing the simulation time for the reduced-order coupled model, we observe that the dynamic iteration is only slightly faster than the monolithic method. This can be explained by the fact that the both circuit and reduced MQS subsystems have very small dimensions.

In Figure 4.22, we present the internal outputs i_M and $-v_M$ of the MQS and circuit subsystems, respectively, as well as the external output $-i_V$ of the coupled system for all four simulation approaches. Taking the solution obtained by the monolithic approach without model reduction as the reference solution, we computed the relative errors in the outputs for remaining three simulation approaches. They are presented in Figure 4.23. One can see in Figure 4.23(a) that the MOR error for the MQS system is smaller than that obtained by dynamic iteration, and the errors accumulate in dynamic iteration combined with model reduction. Furthermore, the dynamic iteration error in the internal output of the circuit system appears to dominate the MOR error, see Figure 4.23(b). Finally, Figure 4.23(c) shows that the MOR, dynamic iteration and combined errors in the external output of the coupled system are about the same.

The dynamic iteration error can be reduced by taking smaller macro step sizes or by performing more iterates, whereas the MOR error can be reduced by taking larger reduced dimensions. However, this increases the simulation time.

	Simulation time in s	Model reduction time in s
monolithic	2.6427e+03	
dynamic iteration	1.1327e+03	
monolithic + model reduction	0.2702	40.0414
dynamic iteration + model reduction	0.2173	40.0414

Table 4.3: Computation time for the coupled MQS-circuit system

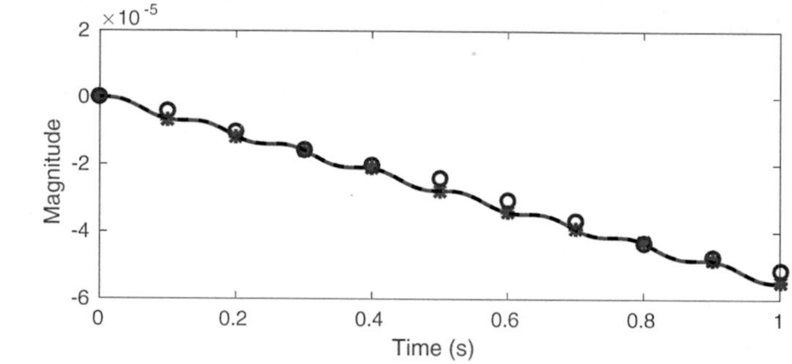

(a) Current i_M

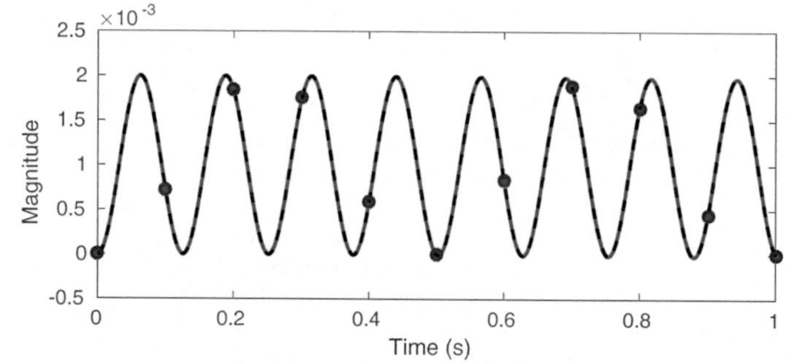

(b) Voltage $-v_M$

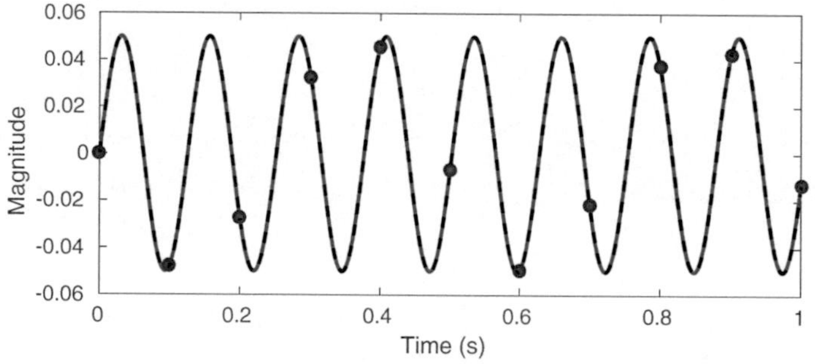

(c) Current $-i_V$

— monolithic approach
• dynamic iteration
- - monolithic approach combined with model reduction
o dynamic iteration combined with model reduction

Figure 4.22: Outputs of the coupled MQS-circuit system

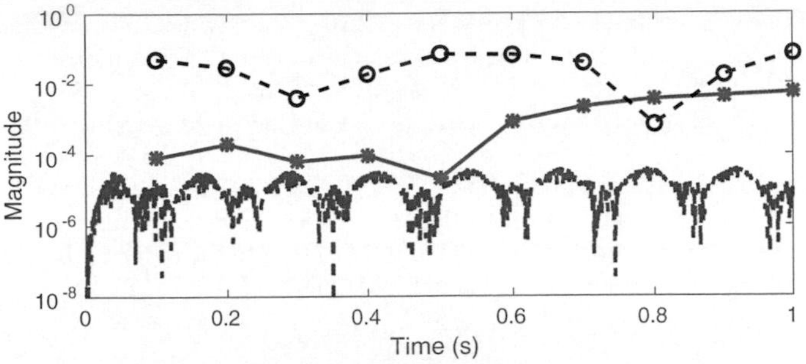

(a) Relative error in current i_M

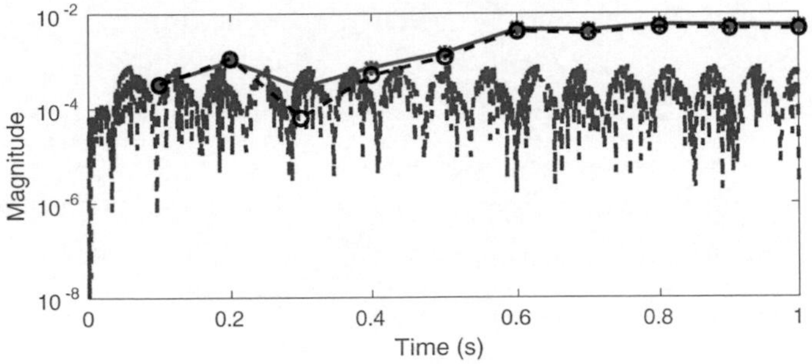

(b) Relative error in voltage $-v_M$

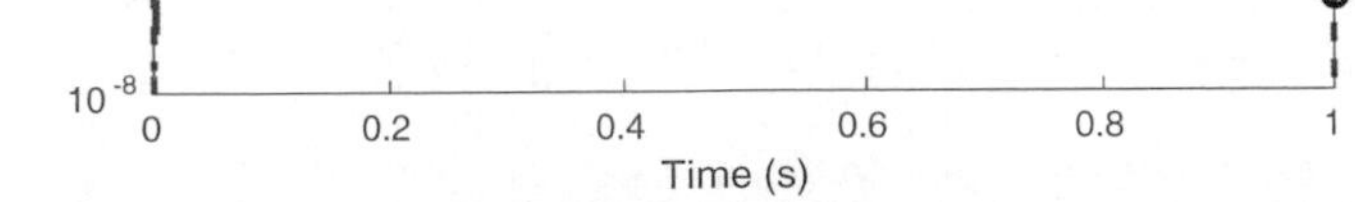

(c) Relative error in current $-i_V$

··•·· dynamic iteration
— monolithic approach combined with model reduction
-o- dynamic iteration combined with model reduction

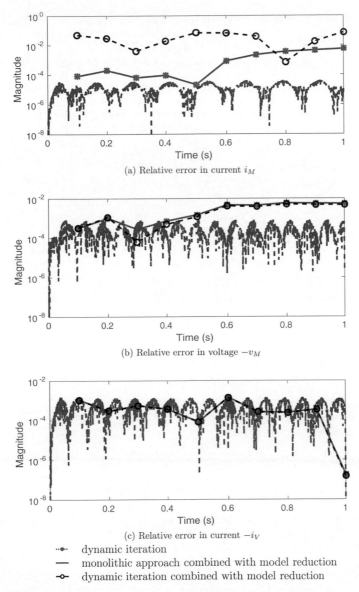

Figure 4.23: Relative errors in the outputs for the coupled MQS-circuit system

4.7 3D Magneto-quasistatic problems

In this section, we study the MQS system (4.2.10) on a 3D domain Ω. First, we consider the FEM discretization of such a system using $H(\mathrm{curl},\Omega)$- conforming finite elements. Then we investigate the DAE structure of the resulting semidiscretized system and present an extension of some results for 2D MQS systems to the 3D case.

We start with introducing appropriate function spaces and deriving a weak formulation for the 3D MQS system (4.2.10). Let $\Omega \subset \mathbb{R}^3$ be a bounded connected domain with a Lipschitz continuous boundary $\partial\Omega$ and let n_0 denote the outer normal of $\partial\Omega$. We define the Sobolev spaces for $\mathbb{R}^3$-valued functions

$$\begin{aligned}
H(\mathrm{div},\Omega) &= \{\varphi \in \mathcal{L}^2(\Omega) : \nabla\cdot\varphi \in \mathcal{L}^2(\Omega)\},\\
H(\mathrm{curl},\Omega) &= \{\varphi \in \mathcal{L}^2(\Omega) : \nabla\times\varphi \in \mathcal{L}^2(\Omega)\},\\
H_0(\mathrm{curl},\Omega) &= \{\varphi \in H(\mathrm{curl},\Omega) : \varphi\times n_0 = 0 \text{ on } \partial\Omega\},\\
G(\Omega) &= \{\psi = \nabla\phi :\ \phi \in H(\mathrm{div},\Omega),\ \phi = c \text{ on } \partial\Omega\},\\
\mathcal{L}^2(\Omega)/G(\Omega) &\simeq \{\varphi \in \mathcal{L}^2(\Omega) : \langle\varphi,\psi\rangle_{\mathcal{L}^2(\Omega)} = 0 \text{ for all } \psi \in G(\Omega)\}.
\end{aligned}$$

On the spaces $H(\mathrm{div},\Omega)$ and $H(\mathrm{curl},\Omega)$, we define the inner products

$$\begin{aligned}
\langle\varphi,\psi\rangle_{H(\mathrm{div},\Omega)} &= \langle\nabla\cdot\varphi,\nabla\cdot\psi\rangle_{\mathcal{L}^2(\Omega)} + \langle\varphi,\psi\rangle_{\mathcal{L}^2(\Omega)},\\
\langle\varphi,\psi\rangle_{H(\mathrm{curl},\Omega)} &= \langle\nabla\times\varphi,\nabla\times\psi\rangle_{\mathcal{L}^2(\Omega)} + \langle\varphi,\psi\rangle_{\mathcal{L}^2(\Omega)},
\end{aligned}$$

respectively, which induce the norms

$$\begin{aligned}
\|\varphi\|_{H(\mathrm{div},\Omega)} &= \left(\|\varphi\|^2_{\mathcal{L}^2(\Omega)} + \|\nabla\cdot\varphi\|^2_{\mathcal{L}^2(\Omega)}\right)^{\frac{1}{2}},\\
\|\varphi\|_{H(\mathrm{curl},\Omega)} &= \left(\|\varphi\|^2_{\mathcal{L}^2(\Omega)} + \|\nabla\times\varphi\|^2_{\mathcal{L}^2(\Omega)}\right)^{\frac{1}{2}}.
\end{aligned}$$

Furthermore, using the definition of $\mathcal{L}^2(0,T;\mathcal{V})$ in Section 2.5.4, we define the space

$$W^{1,2}(0,T;H(\mathrm{curl},\Omega)) = \{\varphi \in \mathcal{L}^2(0,T;H(\mathrm{curl},\Omega)) \ :\ \frac{\partial}{\partial t}\varphi \in \mathcal{L}^2(0,T;(H(\mathrm{curl},\Omega))')\}$$

with the norm

$$\|\varphi\|_{W^{1,2}(0,T;H(\mathrm{curl},\Omega))} = \left(\|\varphi\|^2_{\mathcal{L}^2(0,T;H(\mathrm{curl},\Omega))} + \|\frac{\partial}{\partial t}\varphi\|^2_{\mathcal{L}^2(0,T;H(\mathrm{curl},\Omega))}\right)^{\frac{1}{2}}.$$

We present now a weak formulation for the MQS system (4.2.10). Multiplying the equations with a test function $\psi \in H(\mathrm{curl},\Omega)$ and integrating over Ω, we obtain by using Green's formula the weak formulation

$$\begin{aligned}
\frac{\partial}{\partial t}\int_\Omega \boldsymbol{\sigma}\mathbf{A}\cdot\psi\,d\xi + \int_\Omega \boldsymbol{\nu}(\cdot,\|\nabla\times\mathbf{A}\|)(\nabla\times\mathbf{A})\cdot(\nabla\times\psi)\,d\xi &= \int_\Omega \psi^T\boldsymbol{\chi}_{\mathrm{str}}\iota\,d\xi,\\
\frac{\partial}{\partial t}\int_\Omega \boldsymbol{\chi}_{\mathrm{str}}^T\mathbf{A}\,d\xi + \mathcal{R}\iota &= v,\\
\mathbf{A}(\cdot,0) &= \mathbf{A}_0,
\end{aligned} \tag{4.7.1}$$

with an initial vector $\mathbf{A}_0 \in \mathcal{L}^2(\Omega)\setminus G(\Omega)$.

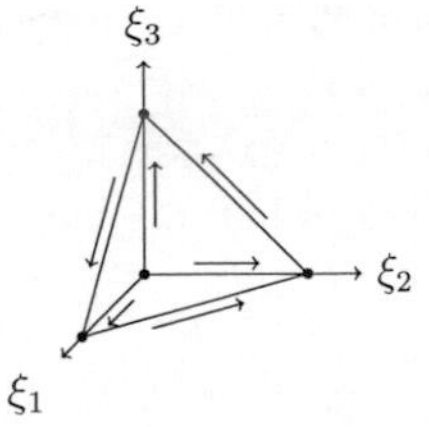

Figure 4.24: Reference configuration $\hat{K}_e$ and degrees of freedom for linear Nédélec elements of first type

4.7.1 Discretization

Let $\mathcal{T}_h(\Omega)$ be a regular simplicial triangulation of Ω, and let n_n, n_e and n_f denote the number of nodes, edges and facets, respectively. For a spatial discretization, we use Nédélec elements of first type as defined in [Néd80]. In the literature, they are also called Whitney elements of first type or edge elements, see [Bos98, Section 5] or [Sch03a]. The reference linear Nédélec element $(\hat{K}_e, \hat{P}_e, \hat{\Sigma}_e)$ of first type is defined by the reference tetrahedron $\hat{K}_e$ with edges $\hat{v}_1, \dots, \hat{v}_6$ as shown in Figure 4.24, the function space

$$\hat{P}_e = \left\langle \begin{bmatrix} 1 \\ 0 \\ 0 \end{bmatrix}, \begin{bmatrix} 0 \\ 1 \\ 0 \end{bmatrix}, \begin{bmatrix} 0 \\ 0 \\ 1 \end{bmatrix}, \begin{bmatrix} 0 \\ \hat{\xi}_3 \\ \hat{\xi}_2 \end{bmatrix}, \begin{bmatrix} \hat{\xi}_3 \\ 0 \\ \hat{\xi}_1 \end{bmatrix}, \begin{bmatrix} \hat{\xi}_2 \\ \hat{\xi}_1 \\ 0 \end{bmatrix} \right\rangle$$

and the six degrees of freedom for $\varphi \in \hat{P}_e$ related to every edge $\hat{v}_i$ given by

$$\hat{\Sigma}_e = \left\{ \varsigma_k(\varphi) = \int_{\hat{v}_k} \hat{t}_k \cdot \varphi \, d\hat{s}, \; k = 1, 2, ..., 6 \right\},$$

where $\hat{t}_k$ denotes the tangential unit vector to the edge $\hat{v}_k$. The reference basis functions for the linear Nédélec element given by

$$\hat{\phi}^e_1(\hat{\xi}) = \begin{bmatrix} 1 - \hat{\xi}_3 - \hat{\xi}_2 \\ \hat{\xi}_1 \\ \hat{\xi}_1 \end{bmatrix}, \quad \hat{\phi}^e_2(\hat{\xi}) = \begin{bmatrix} \hat{\xi}_2 \\ 1 - \hat{\xi}_3 - \hat{\xi}_1 \\ \hat{\xi}_2 \end{bmatrix}, \quad \hat{\phi}^e_3(\hat{\xi}) = \begin{bmatrix} \hat{\xi}_3 \\ \hat{\xi}_3 \\ 1 - \hat{\xi}_2 - \hat{\xi}_1 \end{bmatrix},$$

$$\hat{\phi}^e_4(\hat{\xi}) = \begin{bmatrix} -\hat{\xi}_2 \\ \hat{\xi}_1 \\ 0 \end{bmatrix}, \quad \hat{\phi}^e_5(\hat{\xi}) = \begin{bmatrix} 0 \\ -\hat{\xi}_3 \\ \hat{\xi}_2 \end{bmatrix}, \quad \hat{\phi}^e_6(\hat{\xi}) = \begin{bmatrix} \hat{\xi}_3 \\ 0 \\ -\hat{\xi}_1 \end{bmatrix}$$

are determined by the conditions $\varsigma_k(\hat{\phi}^e_l) = \delta_{k,l}$ with the Kronecker delta $\delta_{k,l}$.

To define the global functions, we use a linear affine mapping

$$F_K(\xi) : \hat{K}_e \to K$$

given by $F_K(\xi) = B_K\xi + b_K$ with a nonsingular matrix $B_K \in \mathbb{R}^{3\times3}$ and a vector $b_K \in \mathbb{R}^3$. To ensure the tangential continuity of the basis functions, we use the Piola transformation defined by

$$\phi^e(\xi) = B_K^{-T}\hat{\phi}^e(F_K^{-1}(\xi)), \quad \xi \in K, \tag{4.7.2}$$

for $\hat{\phi}^e \in \hat{P}_e$. The following lemma shows how to determine $\nabla \times \phi^e$ by applying the curl operator to $\hat{\phi}^e$.

Lemma 4.20 ([Sch03a]). *Let $\hat{K}_e$ be the reference tetrahedron and $K = F_K(\hat{K}_e)$. For a function ϕ^e defined by the Piola transformation (4.7.2) of a reference function $\hat{\phi}^e \in \hat{P}_e$, we have*

$$\nabla \times \phi^e(\xi) = \frac{1}{\det B_K}(\nabla \times \hat{\phi}^e(F_K^{-1}(\xi))), \quad \xi \in K.$$

The global basic functions $\{\phi_j^e\}_{j=1}^{n_e}$ of the finite dimensional space $\mathcal{V}_h \subset H(\mathrm{curl}, \Omega)$ are defined via the Piola transformation (4.7.2) such that the local orientation of the degrees of freedom on two adjacent tetrahedrons matches. In [AV15], the global basis functions are defined as follows

$$\phi_j^e(\xi) = \begin{cases} \left[\mathrm{sign}_{K_n}^k\right] B_{K_n}^T \hat{\phi}_k^e(F_{K_n}^{-1}(\xi)), & \xi \in K_n, \\ \left[\mathrm{sign}_{K_m}^l\right] B_{K_m}^T \hat{\phi}_l^e(F_{K_m}^{-1}(\xi)), & \xi \in K_m, \end{cases}$$

for $K_n, K_m \in \mathcal{T}$, where k and l are the indices of the edges $F_{K_n}^{-1}(v_j)$ and $F_{K_m}^{-1}(v_j)$, respectively, in $\hat{K}_e$, and

$$\begin{aligned}
\left[\mathrm{sign}_{K_n}^k\right] &= +1, & \left[\mathrm{sign}_{K_m}^l\right] &= +1 & \text{if} \quad & \det B_{K_n} > 0, & \det B_{K_m} < 0, \\
\left[\mathrm{sign}_{K_n}^k\right] &= +1, & \left[\mathrm{sign}_{K_m}^l\right] &= -1 & \text{if} \quad & \det B_{K_n} > 0, & \det B_{K_m} > 0, \\
\left[\mathrm{sign}_{K_n}^k\right] &= -1, & \left[\mathrm{sign}_{K_m}^l\right] &= +1 & \text{if} \quad & \det B_{K_n} < 0, & \det B_{K_m} < 0, \\
\left[\mathrm{sign}_{K_n}^k\right] &= -1, & \left[\mathrm{sign}_{K_m}^l\right] &= -1 & \text{if} \quad & \det B_{K_n} < 0, & \det B_{K_m} > 0.
\end{aligned} \tag{4.7.3}$$

This definition applies to all tetrahedra containing the edge v_j. The global function ϕ_j^e is zero everywhere else.

We take the test space $\mathcal{D}(\Omega)$ to be equal to the trial space $\mathcal{V}_h$ and approximate

$$\mathbf{A}(\xi, t) \approx \sum_{j=1}^{n_e} \alpha_j(t)\phi_j^e(\xi).$$

As in the 2D case, we get the nonlinear DAE

$$\begin{bmatrix} \mathcal{M} & 0 \\ \mathcal{X}^T & 0 \end{bmatrix} \frac{d}{dt} \begin{bmatrix} a \\ \iota \end{bmatrix} = \begin{bmatrix} -\mathcal{K}(a) & \mathcal{X} \\ 0 & -\mathcal{R} \end{bmatrix} \begin{bmatrix} a \\ \iota \end{bmatrix} + \begin{bmatrix} 0 \\ I \end{bmatrix} u \tag{4.7.4}$$

with $a = \left[\alpha_1, \ldots, \alpha_{n_e}\right]^T$ and

$$\begin{aligned}
\mathcal{M}_{i,j} &= \textstyle\int_\Omega \boldsymbol{\sigma}\phi_j^e \cdot \phi_i^e \, d\xi, \qquad i, j = 1, \ldots, n_e, \\
\mathcal{K}_{i,j}(a) &= \textstyle\int_\Omega \boldsymbol{\nu}(\cdot, \|\nabla \times \sum_{k=1}^{n_e} \alpha_k \phi_k^e\|) \left(\nabla \times \phi_j^e\right) \cdot \left(\nabla \times \phi_i^e\right) d\xi, \quad i, j = 1, \ldots, n_e, \\
\mathcal{X}_{i,j} &= \textstyle\int_\Omega (\boldsymbol{\chi}_{\mathrm{str}})_j \cdot \phi_i^e \, d\xi, \quad i = 1, \ldots, n_e, \, j = 1, \ldots, m.
\end{aligned} \tag{4.7.5}$$

Here $(\boldsymbol{\chi}_{\mathrm{str}})_j$ denotes the j-th column of $\boldsymbol{\chi}_{\mathrm{str}}$. Note that the matrices $\mathcal{M}$ and $\mathcal{K}(a)$ are symmetric, and, unlike the 2D case, $\mathcal{K}(a)$ is only positive semidefinite. Our goal is now to determine the kernel of $\mathcal{K}(a)$. We introduce Nédélec elements of second type to obtain an alternative representation for $\mathcal{K}(a)$ which allows us to easy determine the kernel of $\mathcal{K}(a)$. These elements are also called face elements, Whitney

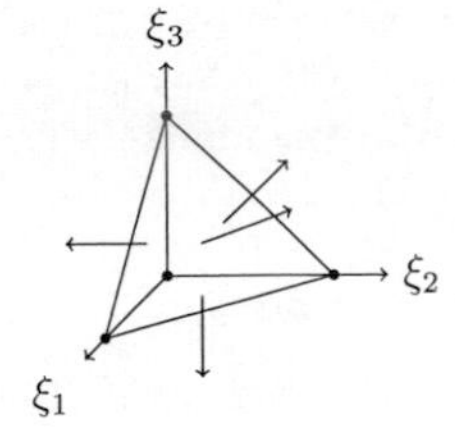

Figure 4.25: Reference configuration $\hat{K}_f$ and degrees of freedom for linear Nédélec elements of second type

elements of second type [Bos98, Section 5] or Raviart-Thomas elements [AV15]. The reference linear Nédélec element of second type is given by $(\hat{K}_f, \hat{P}_f, \hat{\Sigma}_f)$, where $\hat{K}_f$ is the reference thetrahedron with the faces $\hat{f}_1, \ldots, \hat{f}_4$ and degrees of freedom as shown in Figure 4.25. The function space $\hat{P}_f$ is given by

$$\hat{P}_f = \left\langle \begin{bmatrix} 1 \\ 0 \\ 0 \end{bmatrix}, \begin{bmatrix} 0 \\ 1 \\ 0 \end{bmatrix}, \begin{bmatrix} 0 \\ 0 \\ 1 \end{bmatrix}, \begin{bmatrix} \hat{\xi}_1 \\ \hat{\xi}_1 \\ \hat{\xi}_1 \end{bmatrix} \right\rangle,$$

and degrees of freedom are defined as

$$\hat{\Sigma}_f = \left\{ \hat{\varsigma}_k(\hat{u}) = \int_{\hat{f}_k} \hat{\eta}_k \cdot \hat{u}\, ds, \quad k = 1, 2, 3, 4 \right\},$$

where $\hat{\eta}_k$ denotes the outer normal unit vector to the face $\hat{f}_k$. For the function space $\hat{P}_f$, we can use the conditions $\hat{\varsigma}_k(\hat{\phi}^f_l) = \delta_{k,l}$ to specify the reference basis functions

$$\hat{\phi}^f_1(\hat{\xi}) = \begin{bmatrix} \hat{\xi}_1 \\ \hat{\xi}_2 \\ \hat{\xi}_3 - 1 \end{bmatrix}, \quad \hat{\phi}^f_2(\hat{\xi}) = \begin{bmatrix} \hat{\xi}_1 \\ \hat{\xi}_2 - 1 \\ \hat{\xi}_3 \end{bmatrix}, \quad \hat{\phi}^f_3(\hat{\xi}) = \begin{bmatrix} \hat{\xi}_1 - 1 \\ \hat{\xi}_2 \\ 1 - \hat{\xi}_3 \end{bmatrix}, \quad \hat{\phi}^f_4(\hat{\xi}) = \begin{bmatrix} \hat{\xi}_1 \\ \hat{\xi}_2 \\ \hat{\xi}_3 \end{bmatrix}.$$

To ensure the normal continuity of the global basis functions, we again use the affine linear mapping $F_K : \hat{K}_f \to K$ given by $F_K(\hat{\xi}) = B_K\hat{\xi} + b_K$ and define the Piola transformation for Nédélec elements of second type as

$$\phi^f(\xi) = \frac{1}{\det B_K} B_K \hat{\phi}^f(F_K^{-1}(\xi)), \quad \xi \in K,$$

for $\hat{\phi}^f \in \hat{P}_f$. As for Nédélec elements of first type, we also have to ensure the orientation of two adjacent tetrahedra K_n and K_m. Therefore, we define the basis functions ϕ^f_j as

$$\phi^f_j(\xi) = \begin{cases} \left[\text{sign}^k_{K_n}\right] \frac{1}{\det B_{K_n}} B_{K_n} \hat{\phi}^f_k(F_{K_n}^{-1}(\xi)), & \xi \in K_n, \\ \left[\text{sign}^l_{K_m}\right] \frac{1}{\det B_{K_m}} B_{K_m} \hat{\phi}^f_l(F_{K_m}^{-1}(\xi)), & \xi \in K_m, \end{cases}$$

where $\left[\mathrm{sign}^k_{K_n}\right]$ and $\left[\mathrm{sign}^l_{K_m}\right]$ are given in (4.7.3) and k and l are the indices of the faces $F^{-1}_{K_n}(f_j)$ and $F^{-1}_{K_m}(f_j)$, respectively, on the reference tetraheadron $\hat{K}_f$. This definition is applied to all tetrahedra containing the face f_j. For the Nédélec elements of first type $\Phi^e = [\phi^e_1, \ldots, \phi^e_{n_e}]$ and second type $\Phi^f = [\phi^f_1, \ldots, \phi^f_{n_f}]$, we have the following relation

$$\nabla \times \Phi^e = \Phi^f \mathbf{C},$$

where $\mathbf{C} \in \mathbb{R}^{n_f \times n_e}$ is the discrete curl matrix with the entries

$$\mathbf{C}_{ij} = \begin{cases} 1, & \text{if edge } j \text{ belongs to face } i \text{ and their orientations match,} \\ -1, & \text{if edge } j \text{ belongs to face } i \text{ and their orientations do not match,} \\ 0, & \text{if edge } j \text{ does not belong to face } i, \end{cases}$$

see [Bos98, Section 5]. Using this relation, we can rewrite the matrix $\mathcal{K}(a)$ as

$$\begin{aligned} \mathcal{K}(a) &= \int_\Omega \nu(\cdot, \|\nabla \times \Phi^e a\|) \left(\nabla \times \Phi^e\right)^T \left(\nabla \times \Phi^e\right) d\xi \\ &= \int_\Omega \nu(\cdot, \|\Phi^f \mathbf{C} a\|) \mathbf{C}^T (\Phi^f)^T \Phi^f \mathbf{C}\, d\xi \\ &= \mathbf{C}^T \int_\Omega \nu(\cdot, \|\Phi^f \mathbf{C} a\|) (\Phi^f)^T \Phi^f\, d\xi\, \mathbf{C} \\ &= \mathbf{C}^T \mathcal{M}_\nu \left(\mathbf{C} a\right) \mathbf{C}, \end{aligned}$$

where the mass matrix $\mathcal{M}_\nu(\mathbf{C}a)$ is given by

$$(\mathcal{M}_\nu(\mathbf{C}a))_{ij} = \int_\Omega \nu(\cdot, \|\Phi^f \mathbf{C} a\|) \phi^f_j \cdot \phi^f_i\, d\xi, \quad i, j = 1, \ldots, n_f.$$

With this representation, we can now characterize the kernel of $\mathcal{K}(a)$. First, we note that $\mathcal{M}_\nu(a)$ is symmetric, positive definite and

$$\mathrm{rank}(\mathbf{C}) = n_e - n_n + 1,$$

see [Bos98]. We can specify the kernel of matrix $\mathbf{C}$ using the discrete gradient matrix $\mathbf{G}_0 \in \mathbb{R}^{n_e \times n_n}$ defined as

$$(\mathbf{G}_0)_{ij} = \begin{cases} 1, & \text{if edge } i \text{ leaves node } j, \\ -1, & \text{if edge } i \text{ enters node } j, \\ 0, & \text{else.} \end{cases}$$

It holds $\mathbf{C}\mathbf{G}_0 = 0$ and $\mathrm{rank}(\mathbf{G}_0) = n_n - 1$. Then by removing one column of $\mathbf{G}_0$ we get the reduced discrete gradient matrix $\mathbf{G}$, whose columns form a basis of $\ker(\mathbf{C})$ and also of $\ker(\mathcal{K}(a))$.

The coupling matrix $\mathcal{X}$ can also be represented in a factorized form using the discrete curl matrix $\mathbf{C}$. This can be achieved by taking into account the divergence free property of the winding function χ_{str}, which implies $\chi_{\mathrm{str}} = \nabla \times \gamma$ for a certain matrix-valued function $\gamma : \Omega \to \mathbb{R}^{3 \times m}$. In this case, the coupling matrix $\mathcal{X}$ takes the form

$$
\begin{aligned}
\mathcal{X} &= \int_\Omega (\Phi^e)^T \chi_{\text{str}}\, d\xi \\
&= \int_\Omega (\Phi^e)^T \nabla \times \gamma\, d\xi \\
&= \int_\Omega (\nabla \times \Phi^e)^T \gamma\, d\xi \\
&= \mathbf{C}^T \int_\Omega (\Phi^f)^T \gamma\, d\xi \\
&= \mathbf{C}^T \Upsilon,
\end{aligned}
$$

where the entries of $\Upsilon \in \mathbb{R}^{n_f \times m}$ are given by

$$
\Upsilon_{ij} = \int_\Omega \gamma_j \cdot \phi_i^f\, d\xi, \quad i = 1, \ldots, n_f,\, j = 1, \ldots, m,
$$

and γ_j denotes the j-th column of γ. Note that the matrix $\mathcal{X}$ has, as in the 2D case, full column rank because of conditions (4.2.4) and (4.2.5). This implies that Υ has also full column rank.

4.7.2 Regularization of the FEM model

In this section, we study the FEM model (4.7.4). First of all, note that this model is singular since the underlying matrix pencil

$$
\lambda \begin{bmatrix} \mathcal{M} & 0 \\ \mathcal{X}^T & 0 \end{bmatrix} - \begin{bmatrix} -\mathcal{K}(a) & \mathcal{X} \\ 0 & -\mathcal{R} \end{bmatrix} = \lambda \begin{bmatrix} \mathcal{M} & 0 \\ \Upsilon^T \mathbf{C} & 0 \end{bmatrix} - \begin{bmatrix} -\mathbf{C}^T \mathcal{M}_\nu(\mathbf{C}a)\mathbf{C} & \mathbf{C}^T \Upsilon \\ 0 & -\mathcal{R} \end{bmatrix}
$$

is singular. It can be regularized by determining a common kernel of the matrices

$$
\begin{bmatrix} \mathcal{M} & 0 \\ \mathcal{X}^T & 0 \end{bmatrix}, \quad \begin{bmatrix} -\mathcal{K}(a) & \mathcal{X} \\ 0 & -\mathcal{R} \end{bmatrix}.
$$

The DAE system (4.7.4) can be written as

$$
\begin{aligned}
\mathcal{M} \dot{a} &= -\mathcal{K}(a) a + \mathcal{X} \iota, \\
\mathcal{X}^T \dot{a} &= \qquad\qquad -\mathcal{R} \iota + u,
\end{aligned}
$$

where a, $\mathcal{M}$, $\mathcal{K}(a)$ and $\mathcal{X}$ are partitioned into blocks as in (4.3.11) according to the conducting and nonconducting subdomains. Solving the second equation for ι, we obtain

$$
\iota = -\mathcal{R}^{-1} \mathcal{X}^T \dot{a} + \mathcal{R}^{-1} u.
$$

Inserting this vector into the first equation leads to the DAE system

$$
\mathcal{E}_1 \dot{a} \;=\; -\mathcal{K}(a) a + \mathcal{B}_1 u \tag{4.7.6}
$$

with the matrices

$$
\begin{aligned}
\mathcal{E}_1 &= \begin{bmatrix} \mathcal{M}_{11} + \mathcal{X}_1\mathcal{R}^{-1}\mathcal{X}_1^T & \mathcal{X}_1\mathcal{R}^{-1}\mathcal{X}_2^T \\ \mathcal{X}_2\mathcal{R}^{-1}\mathcal{X}_1^T & \mathcal{X}_2\mathcal{R}^{-1}\mathcal{X}_2^T \end{bmatrix}, \\
\mathcal{K}(a) &= \begin{bmatrix} \mathcal{K}_{11}(a_1) & \mathcal{K}_{12} \\ \mathcal{K}_{21} & \mathcal{K}_{22} \end{bmatrix} = \begin{bmatrix} \mathbf{C}_1^T\mathcal{M}_\nu(\mathbf{C}a)\mathbf{C}_1 & \mathbf{C}_1^T\mathcal{M}_\nu(\mathbf{C}a)\mathbf{C}_2 \\ \mathbf{C}_2^T\mathcal{M}_\nu(\mathbf{C}a)\mathbf{C}_1 & \mathbf{C}_2^T\mathcal{M}_\nu(\mathbf{C}a)\mathbf{C}_2 \end{bmatrix}, \\
\mathcal{B}_1 &= \begin{bmatrix} \mathcal{X}_1 \\ \mathcal{X}_2 \end{bmatrix} \mathcal{R}^{-1} = \begin{bmatrix} \mathbf{C}_1^T\Upsilon \\ \mathbf{C}_2^T\Upsilon \end{bmatrix} \mathcal{R}^{-1},
\end{aligned}
\tag{4.7.7}
$$

and $\mathbf{C} = \left[\mathbf{C}_1,\, \mathbf{C}_2\right]$. The output y is given by

$$
y = \iota = -\mathcal{R}^{-1}\left[\mathcal{X}_1^T \quad \mathcal{X}_2^T\right]\dot{a} + \mathcal{R}^{-1}u = -\mathcal{B}_1^T\dot{a} + \mathcal{R}^{-1}u.
$$

The matrix pencil $\lambda\mathcal{E}_1 - \mathcal{K}(a)$ is singular in the sense that $\det(\lambda\mathcal{E}_1 - \mathcal{K}(a)) = 0$ for all $\lambda \in \mathbb{C}$. This follows from the fact that the matrices $\mathcal{E}_1$ und $\mathcal{K}(a)$ have a common kernel. We, therefore, determine $\ker(\mathcal{E}_1) \cap \ker(\mathcal{K}(a))$.

Lemma 4.21. *Assume that $\mathcal{M}_{11}$, $\mathcal{R}$ and $\mathcal{M}_\nu(\mathbf{C}a)$ are positive definite. Let the columns of $Y_{\mathbf{C}_2} \in \mathbb{R}^{n_2\times k_2}$ form a basis of $\ker(\mathbf{C}_2)$. Then $\ker(\mathcal{E}_1) \cap \ker(\mathcal{K}(a))$ is spanned by columns of the matrix $\left[0,\, Y_{\mathbf{C}_2}^T\right]^T$.*

Proof. Using $\mathcal{X} = \mathbf{C}^T\Upsilon$ and (4.7.7), we rewrite the matrices $\mathcal{E}_1$ and $\mathcal{K}(a)$ as

$$
\begin{aligned}
\mathcal{E}_1 &= \begin{bmatrix} I & \mathbf{C}_1^T\Upsilon \\ 0 & \mathbf{C}_2^T\Upsilon \end{bmatrix} \begin{bmatrix} \mathcal{M}_{11} & 0 \\ 0 & \mathcal{R}^{-1} \end{bmatrix} \begin{bmatrix} I & 0 \\ \Upsilon^T\mathbf{C}_1 & \Upsilon^T\mathbf{C}_2 \end{bmatrix}, \\
\mathcal{K}(a) &= \begin{bmatrix} \mathbf{C}_1^T \\ \mathbf{C}_2^T \end{bmatrix} \mathcal{M}_\nu(\mathbf{C}a) \left[\mathbf{C}_1 \quad \mathbf{C}_2\right].
\end{aligned}
$$

Assume $w = \left[w_1^T,\, w_2^T\right]^T \in \ker(\mathcal{E}_1)\cap\ker(\mathcal{K}(a))$. Then due to the positive definiteness of $\mathcal{M}_{11}$ and $\mathcal{R}$, it follows from $w^T\mathcal{E}_1 w = 0$ that

$$
\begin{aligned}
w_1 &= 0, \\
\Upsilon^T(\mathbf{C}_1 w_1 + \mathbf{C}_2 w_2) &= 0.
\end{aligned}
\tag{4.7.8}
$$

Furthermore, from $w^T\mathcal{K}(a)w = 0$ with $w_1 = 0$ and the positive definiteness of $\mathcal{M}_\nu(\mathbf{C}a)$, we get $\mathbf{C}_2 w_2 = 0$. This means that $w_2 \in \ker(\mathbf{C}_2)$ and hence, $w_2 = Y_{\mathbf{C}_2} z$ for some vector z. Note that the vector w_2 also fulfills equation (4.7.8).

Assume now that $w = \left[0 \quad Y_{\mathbf{C}_2}^T\right]^T z$ for some vector z. Then

$$
\mathcal{E}_1 w = \begin{bmatrix} I & \mathbf{C}_1^T\Upsilon \\ 0 & \mathbf{C}_2^T\Upsilon \end{bmatrix} \begin{bmatrix} \mathcal{M}_{11} & 0 \\ 0 & \mathcal{R}^{-1} \end{bmatrix} \begin{bmatrix} I & 0 \\ \Upsilon^T\mathbf{C}_1 & \Upsilon^T\mathbf{C}_2 \end{bmatrix} \begin{bmatrix} 0 \\ Y_{\mathbf{C}_2} \end{bmatrix} z = 0
$$

and

$$
\mathcal{K}(a)w = \begin{bmatrix} \mathbf{C}_1^T \\ \mathbf{C}_2^T \end{bmatrix} \mathcal{M}_\nu(\mathbf{C}a) \left[\mathbf{C}_1 \quad \mathbf{C}_2\right] \begin{bmatrix} 0 \\ Y_{\mathbf{C}_2} \end{bmatrix} z = 0
$$

since $\mathrm{im}(Y_{\mathbf{C}_2}) = \ker(\mathbf{C}_2)$. Therefore, $w \in \ker(\mathcal{E}) \cap \ker(\mathcal{K}(a)$. □

Let the columns of $\hat{Y}_{\mathbf{C}_2} \in \mathbb{R}^{n_2 \times (n_2-k_2)}$ form a basis of $\operatorname{im}(\mathbf{C}_2^T)$. Then the matrix

$$\mathcal{T}_3 = \begin{bmatrix} I & 0 & 0 \\ 0 & \hat{Y}_{\mathbf{C}_2} & Y_{\mathbf{C}_2} \end{bmatrix}$$

is nonsingular. Multiplying system (4.7.6) from the left with $\mathcal{T}_3^T$ and introducing the new state vector

$$\begin{bmatrix} a_1 \\ a_{21} \\ a_{22} \end{bmatrix} = \mathcal{T}_3^{-1} a,$$

the system matrices of the transformed system take the form

$$\mathcal{T}_3^T \mathcal{E}_1 \mathcal{T}_3 = \begin{bmatrix} \mathcal{M}_{11} + \mathcal{X}_1 \mathcal{R}^{-1} \mathcal{X}_1^T & \mathcal{X}_1 \mathcal{R}^{-1} \mathcal{X}_2^T \hat{Y}_{\mathbf{C}_2} & 0 \\ \hat{Y}_{\mathbf{C}_2}^T \mathcal{X}_2 \mathcal{R}^{-1} \mathcal{X}_1^T & \hat{Y}_{\mathbf{C}_2}^T \mathcal{X}_2 \mathcal{R}^{-1} \mathcal{X}_2^T \hat{Y}_{\mathbf{C}_2} & 0 \\ 0 & 0 & 0 \end{bmatrix},$$

$$\mathcal{T}_3^T \mathcal{K}(a) \mathcal{T}_3 = \begin{bmatrix} \mathcal{K}_{11}(a_1) & \mathcal{K}_{12} \hat{Y}_{\mathbf{C}_2} & 0 \\ \hat{Y}_{\mathbf{C}_2}^T \mathcal{K}_{21} & \hat{Y}_{\mathbf{C}_2}^T \mathcal{K}_{22} \hat{Y}_{\mathbf{C}_2} & 0 \\ 0 & 0 & 0 \end{bmatrix},$$

$$\mathcal{T}_3^T \mathcal{B}_1 = \begin{bmatrix} \mathcal{X}_1 \mathcal{R}^{-1} \\ \hat{Y}_{\mathbf{C}_2}^T \mathcal{X}_2 \mathcal{R}^{-1} \\ 0 \end{bmatrix}.$$

This implies that the components of a_{22} are actually not involved in the transformed system. As a consequence, they may be chosen freely. Removing the trivial equation $0 = 0$, we obtain a regular system

$$\mathcal{E}_r \dot{x} = \mathcal{A}_r(x) x + \mathcal{B}_r u, \tag{4.7.9a}$$

$$y = -\mathcal{B}_r^T \dot{x} + \mathcal{R}^{-1} u, \tag{4.7.9b}$$

where $x = \begin{bmatrix} a_1^T, & a_{21}^T \end{bmatrix}^T$ and

$$\begin{aligned} \mathcal{E}_r &= \begin{bmatrix} \mathcal{M}_{11} + \mathcal{X}_1 \mathcal{R}^{-1} \mathcal{X}_1^T & \mathcal{X}_1 \mathcal{R}^{-1} \mathcal{X}_2^T \hat{Y}_{\mathbf{C}_2} \\ \hat{Y}_{\mathbf{C}_2}^T \mathcal{X}_2 \mathcal{R}^{-1} \mathcal{X}_1^T & \hat{Y}_{\mathbf{C}_2}^T \mathcal{X}_2 \mathcal{R}^{-1} \mathcal{X}_2^T \hat{Y}_{\mathbf{C}_2} \end{bmatrix} \in \mathbb{R}^{n_r \times n_r}, \\ \mathcal{A}_r(x) &= \begin{bmatrix} -\mathcal{K}_{11}(a_1) & -\mathcal{K}_{12} \hat{Y}_{\mathbf{C}_2} \\ -\hat{Y}_{\mathbf{C}_2}^T \mathcal{K}_{21} & -\hat{Y}_{\mathbf{C}_2}^T \mathcal{K}_{22} \hat{Y}_{\mathbf{C}_2} \end{bmatrix} \in \mathbb{R}^{n_r \times n_r}, \\ \mathcal{B}_r &= \begin{bmatrix} \mathcal{X}_1 \mathcal{R}^{-1} \\ \hat{Y}_{\mathbf{C}_2}^T \mathcal{X}_2 \mathcal{R}^{-1} \end{bmatrix} \in \mathbb{R}^{n_r \times m} \end{aligned} \tag{4.7.10}$$

with $n_r = n_1 + n_2 - k_2$. Note that the regularity of $\lambda \mathcal{E}_r - \mathcal{A}_r(x)$ follows from the symmetry of $\mathcal{E}_r$ and $\mathcal{A}_r(x)$ and the fact that $\ker(\mathcal{E}_r) \cap \ker(\mathcal{A}_r(x)) = \emptyset$.

Remark 4.22. In order to determine the regularized system (4.7.9), we need the basis matrix $\hat{Y}_{\mathbf{C}_2}$ with $\operatorname{im}(\hat{Y}_{\mathbf{C}_2}) = \operatorname{im}(\mathbf{C}_2^T)$. Such a matrix can be computed using the graph-theoretical algorithm as presented in [Ipa13]. First of all, note that the matrices $\mathbf{C}$ and $\mathbf{G}_0^T$ can be considered as the loop and incidence matrices, respectively, of a directed graph whose nodes and branches correspond to the nodes and edges of the triangulation $\mathcal{T}_h(\Omega)$, see [Deo74]. Let the reduced gradient matrix $\mathbf{G} = \begin{bmatrix}\mathbf{G}_1^T & \mathbf{G}_2^T\end{bmatrix}^T$ be partitioned into blocks according to $\mathbf{C} = \begin{bmatrix}\mathbf{C}_1 & \mathbf{C}_2\end{bmatrix}$. It follows from [Ipa13, Theorem 9] that

$$\ker(\mathbf{C}_2) = \operatorname{im}(\mathbf{G}_2 Z_1),$$

where the columns of Z_1 form a basis of $\ker(\mathbf{G}_1)$. Then $\hat{Y}_{\mathbf{C}_2}$ can be determined as a basis of $\ker(Z_1^T\mathbf{G}_2^T)$. The basis Z_1 can be computed by using the function `kernelAT` from [Ipa13, Section 3] and applying it to $\mathbf{G}_1^T$. The basis $\hat{Y}_{\mathbf{C}_2}$ can be determined as $\hat{Y}_{\mathbf{C}_2} = \texttt{kernelAk}(Z_1^T\mathbf{G}_2^T)$ with the function `kernelAk` from [Ipa13, Section 4.2].

Next, we investigate the tractability index of the regularized 3D MQS system (4.7.9a), (4.7.10).

Theorem 4.23. *Consider a DAE* (4.7.9a), *where* $\mathcal{M}_{11}$ *and* $\mathcal{M}_\nu(\mathbf{C}a)$ *are symmetric, positive definite,* $\mathcal{X}_2$ *has full column rank, and* $\hat{Y}_{\mathbf{C}_2}$ *is a basis of* $\operatorname{im}(\mathbf{C}_2^T)$. *This system has tractability index one.*

Proof. Let the columns of $\mathcal{Y}$ form an orthonormal basis of $\ker(\mathcal{X}_2^T\hat{Y}_{\mathbf{C}_2})$. Then a projector $\mathcal{Q}_0$ onto $\ker(\mathcal{G}_0)$ with $\mathcal{G}_0 = \mathcal{E}_r$ can be defined as

$$\mathcal{Q}_0 = \begin{bmatrix} 0 & 0 \\ 0 & \mathcal{Y}\mathcal{Y}^T \end{bmatrix}.$$

In this case, for $\mathcal{B}_0(x) = -\frac{\partial}{\partial x}(\mathcal{A}_r(x)x)$, the matrix

$$\begin{aligned}\mathcal{G}_1 &= \mathcal{G}_0 + \mathcal{B}_0(x)\mathcal{Q}_0 \\ &= \begin{bmatrix} \mathcal{M}_{11} + \mathcal{X}_1\mathcal{R}^{-1}\mathcal{X}_1^T & \mathcal{X}_1\mathcal{R}^{-1}\mathcal{X}_2^T\hat{Y}_{\mathbf{C}_2} + \mathcal{K}_{12}\hat{Y}_{\mathbf{C}_2}\mathcal{Y}\mathcal{Y}^T \\ \hat{Y}_{\mathbf{C}_2}^T\mathcal{X}_2\mathcal{R}^{-1}\mathcal{X}_1^T & \hat{Y}_{\mathbf{C}_2}^T\mathcal{X}_2\mathcal{R}^{-1}\mathcal{X}_2^T\hat{Y}_{\mathbf{C}_2} + \hat{Y}_{\mathbf{C}_2}^T\mathcal{K}_{22}\hat{Y}_{\mathbf{C}_2}\mathcal{Y}\mathcal{Y}^T \end{bmatrix}\end{aligned}$$

is independent of x. To show that $\mathcal{G}_1$ is invertible, we consider the equation

$$\mathcal{G}_1[v_1, v_2]^T = 0,$$

which is equivalent to

$$(\mathcal{M}_{11} + \mathcal{X}_1\mathcal{R}^{-1}\mathcal{X}_1^T)v_1 + (\mathcal{X}_1\mathcal{R}^{-1}\mathcal{X}_2^T\hat{Y}_{\mathbf{C}_2} + \mathcal{K}_{12}\hat{Y}_{\mathbf{C}_2}\mathcal{Y}\mathcal{Y}^T)v_2 = 0, \tag{4.7.11}$$

$$\hat{Y}_{\mathbf{C}_2}^T\mathcal{X}_2\mathcal{R}^{-1}\mathcal{X}_1^T v_1 + (\hat{Y}_{\mathbf{C}_2}^T\mathcal{X}_2\mathcal{R}^{-1}\mathcal{X}_2^T\hat{Y}_{\mathbf{C}_2} + \hat{Y}_{\mathbf{C}_2}^T\mathcal{K}_{22}\hat{Y}_{\mathbf{C}_2}\mathcal{Y}\mathcal{Y}^T)v_2 = 0. \tag{4.7.12}$$

Multiplying equation (4.7.12) from the left with $\mathcal{Y}^T$ and using $\mathcal{Y}^T\hat{Y}_{\mathbf{C}_2}^T\mathcal{X}_2 = 0$, we obtain $\mathcal{Y}^T\hat{Y}_{\mathbf{C}_2}^T\mathcal{K}_{22}\hat{Y}_{\mathbf{C}_2}\mathcal{Y}\mathcal{Y}^T v_2 = 0$. Since $\mathcal{M}_\nu(\mathbf{C}a)$ is positive definite and $\hat{Y}_{\mathbf{C}_2}$ is a basis of $\operatorname{im}(\mathbf{C}_2^T)$, $\hat{Y}_{\mathbf{C}_2}^T\mathcal{K}_{22}\hat{Y}_{\mathbf{C}_2} = \hat{Y}_{\mathbf{C}_2}^T\mathbf{C}_2^T\mathcal{M}_\nu(\mathbf{C}a)\mathbf{C}_2\hat{Y}_{\mathbf{C}_2}$ is symmetric, positive definite, and, hence, $\mathcal{Y}\mathcal{Y}^T v_2 = 0$. Using the fact that $\mathcal{Y}$ has full column rank, we obtain $\mathcal{Y}^T v_2 = 0$.

Next, we show that $\hat{Y}_{\mathbf{C}_2}^T\mathcal{X}_2$ has full column rank, and, hence the matrix $\mathcal{X}_2^T\hat{Y}_{\mathbf{C}_2}\hat{Y}_{\mathbf{C}_2}^T\mathcal{X}_2$ is invertible. Indeed, let $\hat{Y}_{\mathbf{C}_2}^T\mathcal{X}_2 w = 0$. Then $\mathcal{X}_2 w \in \ker(\hat{Y}_{\mathbf{C}_2}^T)$. On the other side, $\mathcal{X}_2 w = \mathbf{C}_2^T\Upsilon w \in \operatorname{im}(\mathbf{C}_2^T) = \operatorname{im}(\hat{Y}_{\mathbf{C}_2})$. Therefore, $\mathcal{X}_2 w = 0$. Since $\mathcal{X}_2$ has full column rank, we get $w = 0$.

Multiplying (4.7.12) from the left with $\mathcal{X}_1(\mathcal{X}_2^T\hat{Y}_{\mathbf{C}_2}\hat{Y}_{\mathbf{C}_2}^T\mathcal{X}_2)^{-1}\mathcal{X}_2^T\hat{Y}_{\mathbf{C}_2}$ and inserting $\mathcal{Y}^T v_2 = 0$ yields

$$\mathcal{X}_1\mathcal{R}^{-1}\mathcal{X}_1^T v_1 + \mathcal{X}_1\mathcal{R}^{-1}\mathcal{X}_2^T\hat{Y}_{\mathbf{C}_2}v_2 = 0. \tag{4.7.13}$$

Subtracting equation (4.7.13) from (4.7.11) and inserting $\mathcal{Y}^T v_2 = 0$ implies

$$\mathcal{M}_{11}v_1 = 0.$$

Since $\mathcal{M}_{11}$ is invertible, we obtain $v_1 = 0$. Furthermore, multiplying (4.7.12) from left with v_2^T and using $\mathcal{Y}^T v_2 = 0$ and $v_1 = 0$ leads to $v_2^T\hat{Y}_{\mathbf{C}_2}^T\mathcal{X}_2\mathcal{R}^{-1}\mathcal{X}_2^T\hat{Y}_{\mathbf{C}_2}v_2 = 0$. Since $\mathcal{R}$ is positive definite, v_2 is in $\ker(\mathcal{X}_2^T\hat{Y}_{\mathbf{C}_2}) = \mathrm{im}(\mathcal{Y})$. This means that v_2 belongs to the image of $\mathcal{Y}$ and to the kernel of $\mathcal{Y}^T$. Therefore, $v_2 = 0$. Thus, $\mathcal{G}_1$ is invertible, and, hence, by Definition 2.3, system (4.7.9a) is of tractability index 1. □

We consider now the output equation (4.7.9b). Our goal is to transform this equation to the standard form $y = \mathcal{C}_r x$ with an output matrix $\mathcal{C}_r \in \mathbb{R}^{m\times n_r}$. For this purpose, we rewrite the system matrices in (4.7.10) in the short form

$$\mathcal{E}_r = \mathcal{F}_\sigma\mathcal{M}_\sigma\mathcal{F}_\sigma^T, \qquad \mathcal{A}_r(x) = -\mathcal{F}_\nu\mathcal{M}_\nu(\mathcal{F}_\nu^T x)\mathcal{F}_\nu^T, \tag{4.7.14}$$

$$\mathcal{B}_r = \mathcal{F}_\nu\Upsilon\mathcal{R}^{-1} = \mathcal{F}_\sigma\mathcal{M}_\sigma\begin{bmatrix}0\\ I\end{bmatrix}, \tag{4.7.15}$$

where

$$\mathcal{F}_\sigma = \begin{bmatrix} I & \mathcal{X}_1 \\ 0 & \hat{Y}_{\mathbf{C}_2}^T\mathcal{X}_2\end{bmatrix} = \begin{bmatrix} I & \mathbf{C}_1^T\Upsilon \\ 0 & \hat{Y}_{\mathbf{C}_2}^T\mathbf{C}_2^T\Upsilon\end{bmatrix}, \qquad \mathcal{M}_\sigma = \begin{bmatrix}\mathcal{M}_{11} & 0 \\ 0 & \mathcal{R}^{-1}\end{bmatrix}, \qquad \mathcal{F}_\nu = \begin{bmatrix}\mathbf{C}_1^T \\ \hat{Y}_{\mathbf{C}_2}^T\mathbf{C}_2^T\end{bmatrix}.$$

This shows once again that $\mathcal{E}_r$ is positive semidefinite and $\mathcal{A}_r(x)$ is negative semidefinite, since $\mathcal{M}_{11}$, $\mathcal{R}$ and $\mathcal{M}_\nu(\mathcal{F}_\nu^T x)$ are positive definite.

The following theorem provides a condensed form for the pencil $\lambda\mathcal{E}_r - \mathcal{A}_r(x)$ which allows us to extract the algebraic constraints in (4.7.9a) and derive the output matrix $\mathcal{C}_r$.

Theorem 4.24. *Let the matrices $\mathcal{E}_r$ and $\mathcal{A}_r(x)$ be as in (4.7.14). Then there exists a nonsingular matrix $\mathcal{W}$ such that*

$$\mathcal{W}^T\mathcal{E}_r\mathcal{W} = \begin{bmatrix}\mathcal{E}_{11} & 0 & 0\\ 0 & I & 0\\ 0 & 0 & 0\end{bmatrix}, \qquad \mathcal{W}^T\mathcal{A}_r(x)\mathcal{W} = \begin{bmatrix}\mathcal{A}_{11}(x) & 0 & 0\\ 0 & 0 & 0\\ 0 & 0 & I\end{bmatrix}, \tag{4.7.16}$$

where $\mathcal{E}_{11}$ and $-\mathcal{A}_{11}(x)$ are both symmetric and positive definite.

Proof. Let the columns of $\mathcal{Y}_\sigma$ and $\mathcal{Y}_\nu$ form the basis of $\ker(\mathcal{F}_\sigma^T)$ and $\ker(\mathcal{F}_\nu^T)$, respectively, i.e.,

$$\mathrm{im}(\mathcal{Y}_\sigma) = \ker(\mathcal{F}_\sigma^T), \quad \mathrm{im}(\mathcal{Y}_\nu) = \ker(\mathcal{F}_\nu^T). \tag{4.7.17}$$

First, note that $\mathcal{Y}_\sigma^T\mathcal{A}_r(x)$ is independent of x. This follows from the fact that the basis matrix $\mathcal{Y}_\sigma$ has the form $\left[0,\, \mathcal{Y}_2^T\right]^T$, where the columns of $\mathcal{Y}_2$ form the basis of $\ker(\mathcal{X}_2^T\hat{Y}_{\mathbf{C}_2})$. In this case, we have

$$\mathcal{Y}_\sigma^T\mathcal{A}_r(x) = \begin{bmatrix}0 & \mathcal{Y}_2^T\end{bmatrix}\begin{bmatrix}-\mathcal{K}_{11}(a_1) & -\mathcal{K}_{12}\hat{Y}_{\mathbf{C}_2}\\ -\hat{Y}_{\mathbf{C}_2}^T\mathcal{K}_{21} & -\hat{Y}_{\mathbf{C}_2}^T\mathcal{K}_{22}\hat{Y}_{\mathbf{C}_2}\end{bmatrix} = -\mathcal{Y}_2^T\begin{bmatrix}\hat{Y}_{\mathbf{C}_2}^T\mathcal{K}_{21} & \hat{Y}_{\mathbf{C}_2}^T\mathcal{K}_{22}\hat{Y}_{\mathbf{C}_2}\end{bmatrix},$$

which is independent of x. Therefore, in the following, we will just write $\mathcal{Y}_\sigma^T\mathcal{A}_r$.

Due to $\ker(\mathcal{E}_r)\cap\ker(\mathcal{A}_r(x))=\emptyset$, the matrices $\mathcal{Y}_\nu^T\mathcal{E}_r\mathcal{Y}_\nu$ and $\mathcal{Y}_\sigma^T\mathcal{A}_r\mathcal{Y}_\sigma$ are both nonsingular and $\mathcal{Y}_\sigma^T\mathcal{Y}_\nu=0$. Furthermore, let the matrix $\mathcal{W}_1$ be chosen such that the matrix

$$\widehat{\mathcal{W}}=\begin{bmatrix}\mathcal{W}_1 & \mathcal{Y}_\nu(\mathcal{Y}_\nu^T\mathcal{E}_r\mathcal{Y}_\nu)^{-\frac{1}{2}} & \mathcal{Y}_\sigma(\mathcal{Y}_\sigma^T\mathcal{A}_r\mathcal{Y}_\sigma)^{-\frac{1}{2}}\end{bmatrix}$$

is nonsingular. Taking into account (4.7.17), we have

$$\widehat{\mathcal{W}}^T\mathcal{E}_r\widehat{\mathcal{W}}=\begin{bmatrix}\mathcal{W}_1^T\mathcal{E}_r\mathcal{W}_1 & \mathcal{W}_1^T\mathcal{E}_r\mathcal{Y}_\nu(\mathcal{Y}_\nu^T\mathcal{E}_r\mathcal{Y}_\nu)^{-\frac{1}{2}} & 0\\ (\mathcal{Y}_\nu^T\mathcal{E}_r\mathcal{Y}_\nu)^{-\frac{1}{2}}\mathcal{Y}_\nu^T\mathcal{E}_r\mathcal{W}_1 & I_{n_\nu} & 0\\ 0 & 0 & 0\end{bmatrix}=:\hat{\mathcal{E}}_r,$$

$$\widehat{\mathcal{W}}^T\mathcal{A}_r(x)\widehat{\mathcal{W}}=\begin{bmatrix}\mathcal{W}_1^T\mathcal{A}_r(x)\mathcal{W}_1 & 0 & \mathcal{W}_1^T\mathcal{A}_r\mathcal{Y}_\sigma(\mathcal{Y}_\sigma^T\mathcal{A}_r\mathcal{Y}_\sigma)^{-\frac{1}{2}}\\ 0 & 0 & 0\\ (\mathcal{Y}_\sigma^T\mathcal{A}_r\mathcal{Y}_\sigma)^{-\frac{1}{2}}\mathcal{Y}_\sigma^T\mathcal{A}_r\mathcal{W}_1 & 0 & I_{n_\sigma}\end{bmatrix}=:\hat{\mathcal{A}}_r(x),$$

where $n_\nu=\dim(\ker(\mathcal{F}_\nu^T))=\dim(\ker(\mathcal{A}_r(x)))$, $n_\sigma=\dim(\ker(\mathcal{F}_\sigma^T))=\dim(\ker(\mathcal{E}_r))$. The off-diagonal blocks in $\hat{\mathcal{E}}_r$ and $\hat{\mathcal{A}}_r(x)$ can be eliminated by the transformation matrix

$$\widetilde{\mathcal{W}}=\begin{bmatrix}I_{n_r-n_\nu-n_\sigma} & 0 & 0\\ -(\mathcal{Y}_\nu^T\mathcal{E}_r\mathcal{Y}_\nu)^{-\frac{1}{2}}\mathcal{Y}_\nu^T\mathcal{E}_r\mathcal{W}_1 & I_{n_\nu} & 0\\ -(\mathcal{Y}_\sigma^T\mathcal{A}_r\mathcal{Y}_\sigma)^{-\frac{1}{2}}\mathcal{Y}_\sigma^T\mathcal{A}_r\mathcal{W}_1 & 0 & I_{n_\sigma}\end{bmatrix}$$

which is independent of x. Thus, for

$$\mathcal{W}=\widehat{\mathcal{W}}\widetilde{\mathcal{W}}=\begin{bmatrix}\Pi\mathcal{W}_1 & \mathcal{Y}_\nu(\mathcal{Y}_\nu^T\mathcal{E}_r\mathcal{Y}_\nu)^{-\frac{1}{2}} & \mathcal{Y}_\sigma(\mathcal{Y}_\sigma^T\mathcal{A}_r\mathcal{Y}_\sigma)^{-\frac{1}{2}}\end{bmatrix}\tag{4.7.18}$$

with the projector

$$\Pi=I-\mathcal{Y}_\nu(\mathcal{Y}_\nu^T\mathcal{E}_r\mathcal{Y}_\nu)^{-1}\mathcal{Y}_\nu^T\mathcal{E}_r-\mathcal{Y}_\sigma(\mathcal{Y}_\sigma^T\mathcal{A}_r\mathcal{Y}_\sigma)^{-1}\mathcal{Y}_\sigma^T\mathcal{A}_r,\tag{4.7.19}$$

we obtain (4.7.16), where

$$\mathcal{E}_{11}=\mathcal{W}_1^T\Pi^T\mathcal{E}_r\Pi\mathcal{W}_1,\qquad \mathcal{A}_{11}(x)=\mathcal{W}_1^T\Pi^T\mathcal{A}_r(x)\Pi\mathcal{W}_1$$

are symmetric and $\mathcal{E}_{11}$ and $-\mathcal{A}_{11}(x)$ are positive definite. The last observation immediately follows from the positive semidefiniteness of $\mathcal{E}_r$ and $-\mathcal{A}_r(x)$ and the relations

$$\begin{aligned}\operatorname{rank}(\mathcal{W}^T\mathcal{E}_r\mathcal{W})&=\operatorname{rank}(\mathcal{E}_r)=n_r-n_\sigma,\\ \operatorname{rank}(\mathcal{W}^T\mathcal{A}_r(x)\mathcal{W})&=\operatorname{rank}(\mathcal{A}_r(x))=n_r-n_\nu.\end{aligned}$$

This completes the proof. □

For the transformation matrix $\mathcal{W}$ as in (4.7.18), we calculate the inverse $\mathcal{W}^{-1}=\begin{bmatrix}\widehat{\mathcal{W}}_1^T & \widehat{\mathcal{W}}_2^T & \widehat{\mathcal{W}}_3^T\end{bmatrix}^T$ which satisfies

$$I=\mathcal{W}^{-1}\mathcal{W}=\begin{bmatrix}\widehat{\mathcal{W}}_1\Pi\mathcal{W}_1 & \widehat{\mathcal{W}}_1\mathcal{Y}_\nu(\mathcal{Y}_\nu^T\mathcal{E}_r\mathcal{Y}_\nu)^{-\frac{1}{2}} & \widehat{\mathcal{W}}_1\mathcal{Y}_\sigma(\mathcal{Y}_\sigma^T\mathcal{A}_r\mathcal{Y}_\sigma)^{-\frac{1}{2}}\\ \widehat{\mathcal{W}}_2\Pi\mathcal{W}_1 & \widehat{\mathcal{W}}_2\mathcal{Y}_\nu(\mathcal{Y}_\nu^T\mathcal{E}_r\mathcal{Y}_\nu)^{-\frac{1}{2}} & \widehat{\mathcal{W}}_2\mathcal{Y}_\sigma(\mathcal{Y}_\sigma^T\mathcal{A}_r\mathcal{Y}_\sigma)^{-\frac{1}{2}}\\ \widehat{\mathcal{W}}_3\Pi\mathcal{W}_1 & \widehat{\mathcal{W}}_3\mathcal{Y}_\nu(\mathcal{Y}_\nu^T\mathcal{E}_r\mathcal{Y}_\nu)^{-\frac{1}{2}} & \widehat{\mathcal{W}}_3\mathcal{Y}_\sigma(\mathcal{Y}_\sigma^T\mathcal{A}_r\mathcal{Y}_\sigma)^{-\frac{1}{2}}\end{bmatrix}.$$

This equation gives

$$\begin{aligned}\widehat{\mathcal{W}}_2 &= (\mathcal{Y}_\nu^T \mathcal{E}_r \mathcal{Y}_\nu)^{-\frac{1}{2}} \mathcal{Y}_\nu^T \mathcal{E}_r,\\ \widehat{\mathcal{W}}_3 &= (\mathcal{Y}_\sigma^T \mathcal{A}_r \mathcal{Y}_\sigma)^{-\frac{1}{2}} \mathcal{Y}_\sigma^T \mathcal{A}_r\end{aligned}$$

and the conditions $\widehat{\mathcal{W}}_1 \Pi \mathcal{W}_1 = I$, $\widehat{\mathcal{W}}_1 \mathcal{Y}_\nu = 0$, and $\widehat{\mathcal{W}}_1 \mathcal{Y}_\sigma = 0$ for $\widehat{\mathcal{W}}_1$.

We can define now the projectors

$$\Pi_0 = \mathcal{W} \begin{bmatrix} 0 & & \\ & I & \\ & & 0 \end{bmatrix} \mathcal{W}^{-1} = \mathcal{Y}_\nu (\mathcal{Y}_\nu^T \mathcal{E}_r \mathcal{Y}_\nu)^{-1} \mathcal{Y}_\nu^T \mathcal{E}_r,$$

$$\Pi_\infty = \mathcal{W} \begin{bmatrix} 0 & & \\ & 0 & \\ & & I \end{bmatrix} \mathcal{W}^{-1} = \mathcal{Y}_\sigma (\mathcal{Y}_\sigma^T \mathcal{A}_r \mathcal{Y}_\sigma)^{-1} \mathcal{Y}_\sigma^T \mathcal{A}_r$$

onto the right deflating subspaces of $\lambda \mathcal{E}_r - \mathcal{A}_r(x)$ corresponding to the zero and infinite eigenvalues. Note that these projectors and the projector Π in (4.7.19) satisfy the relation $\Pi + \Pi_0 + \Pi_\infty = I$. Furthermore, we introduce the pseudoinverse of $\mathcal{E}_r$ given by

$$\mathcal{E}_r^- = \mathcal{W} \begin{bmatrix} \mathcal{E}_{11}^{-1} & & \\ & I & \\ & & 0 \end{bmatrix} \mathcal{W}^T. \tag{4.7.20}$$

Simple calculations show that this matrix satisfies

$$(\mathcal{E}_r^-)^T = \mathcal{E}_r^-, \tag{4.7.21}$$
$$\mathcal{E}_r \mathcal{E}_r^- \mathcal{E}_r = \mathcal{E}_r, \tag{4.7.22}$$
$$\mathcal{E}_r^- \mathcal{E}_r \mathcal{E}_r^- = \mathcal{E}_r^-, \tag{4.7.23}$$
$$\mathcal{E}_r^- \mathcal{E}_r = I - \Pi_\infty, \tag{4.7.24}$$
$$\mathcal{E}_r \mathcal{E}_r^- = I - \Pi_\infty^T, \tag{4.7.25}$$
$$\mathcal{E}_r \mathcal{E}_r^- \mathcal{A}_r(x) = \mathcal{A}_r(x) \mathcal{E}_r^- \mathcal{E}_r = \mathcal{E}_r \mathcal{E}_r^- \mathcal{A}_r(x) \mathcal{E}_r^- \mathcal{E}_r. \tag{4.7.26}$$

Equations (4.7.21), (4.7.22) and (4.7.23) imply that $\mathcal{E}_r^-$ is the symmetric reflexive inverse of $\mathcal{E}_r$. Note that the input matrix $\mathcal{B}_r$ in (4.7.15) can also be presented as

$$\begin{aligned}\mathcal{B}_r = \mathcal{F}_\sigma \mathcal{M}_\sigma \begin{bmatrix} 0 \\ I \end{bmatrix} &= \mathcal{F}_\sigma \mathcal{M}_\sigma \begin{bmatrix} I & 0 \\ X_1^T & X_2^T \hat{Y}_{\mathbf{C}_2} \end{bmatrix} \begin{bmatrix} 0 \\ \hat{Y}_{\mathbf{C}_2}^T X_2 (X_2^T \hat{Y}_{\mathbf{C}_2} \hat{Y}_{\mathbf{C}_2}^T X_2)^{-1} \end{bmatrix}\\ &= \mathcal{F}_\sigma \mathcal{M}_\sigma \mathcal{F}_\sigma^T \begin{bmatrix} 0 \\ Z \end{bmatrix} = \mathcal{E}_r \begin{bmatrix} 0 \\ Z \end{bmatrix}\end{aligned} \tag{4.7.27}$$

with $Z = \hat{Y}_{\mathbf{C}_2}^T X_2 (X_2^T \hat{Y}_{\mathbf{C}_2} \hat{Y}_{\mathbf{C}_2}^T X_2)^{-1}$. Then using (4.7.22) and the state equation (4.7.9a), the output in (4.7.9b) can be written as

$$\begin{aligned}y &= -\mathcal{B}_r^T \dot{x} + \mathcal{R}^{-1} u\\ &= -\begin{bmatrix} 0 & Z^T \end{bmatrix} \mathcal{E}_r \dot{x} + \mathcal{R}^{-1} u\\ &= -\begin{bmatrix} 0 & Z^T \end{bmatrix} \mathcal{E}_r \mathcal{E}_r^- \mathcal{E}_r \dot{x} + \mathcal{R}^{-1} u\\ &= -\mathcal{B}_r^T \mathcal{E}_r^- (\mathcal{A}_r(x) x + \mathcal{B}_r u) + \mathcal{R}^{-1} u\\ &= -\mathcal{B}_r^T \mathcal{E}_r^- \mathcal{A}_r(x) x + (\mathcal{R}^{-1} - \mathcal{B}_r^T \mathcal{E}_r^- \mathcal{B}_r) u.\end{aligned}$$

Taking into account the special block structure of $\mathcal{A}_r(x)$ in (4.7.10) and using equations (4.7.25) and (4.7.26), we obtain that the matrix

$$\begin{aligned}\mathcal{C}_r &:= -\mathcal{B}_r^T \mathcal{E}_r^- \mathcal{A}_r(x) = -\begin{bmatrix}0 & Z^T\end{bmatrix}(I-\Pi_\infty^T)\mathcal{A}_r(x) = -\begin{bmatrix}0 & Z^T\end{bmatrix}\mathcal{A}_r(x)(I-\Pi_\infty)\\ &= (\mathcal{X}_2^T \hat{Y}_{\mathbf{C}_2}\hat{Y}_{\mathbf{C}_2}^T \mathcal{X}_2)^{-1}\mathcal{X}_2^T \hat{Y}_{\mathbf{C}_2}\hat{Y}_{\mathbf{C}_2}^T \begin{bmatrix}\mathcal{K}_{21} & \mathcal{K}_{22}\hat{Y}_{\mathbf{C}_2}\end{bmatrix}(I-\Pi_\infty) \end{aligned} \qquad (4.7.28)$$

is independent of x. Moreover, it follows from (4.7.22) and (4.7.27) that

$$\begin{aligned}\mathcal{B}_r^T \mathcal{E}_r^- \mathcal{B}_r &= \begin{bmatrix}0 & Z^T\end{bmatrix}\mathcal{E}_r\mathcal{E}_r^-\mathcal{E}_r\begin{bmatrix}0\\ Z\end{bmatrix} = \begin{bmatrix}0 & Z^T\end{bmatrix}\mathcal{F}_\sigma\mathcal{M}_\sigma\mathcal{F}_\sigma^T\begin{bmatrix}0\\ Z\end{bmatrix}\\ &= \begin{bmatrix}0 & I\end{bmatrix}\mathcal{M}_\sigma\begin{bmatrix}0\\ I\end{bmatrix} = \mathcal{R}^{-1},\end{aligned}$$

and, hence, the output takes the form $y = \mathcal{C}_r x$ with $\mathcal{C}_r$ as in (4.7.28). Thus, the regularized DAE system is given by

$$\begin{aligned}\mathcal{E}_r\dot{x} &= \mathcal{A}_r(x)x + \mathcal{B}_r u,\\ y &= \mathcal{C}_r x.\end{aligned} \qquad (4.7.29)$$

For the transformation matrix $\mathcal{W}$ as in (4.7.18), we obtain using (4.7.15) that

$$\mathcal{W}^T\mathcal{B}_r = \begin{bmatrix}\mathcal{W}_1^T\Pi^T\mathcal{B}_r\\ (\mathcal{Y}_\nu^T\mathcal{E}_r\mathcal{Y}_\nu)^{-\frac{1}{2}}\mathcal{Y}_\nu^T\mathcal{F}_\nu\Upsilon\mathcal{R}^{-1}\\ (\mathcal{Y}_\sigma^T\mathcal{A}_r\mathcal{Y}_\sigma)^{-\frac{1}{2}}\mathcal{Y}_\sigma^T\mathcal{F}_\sigma\mathcal{M}_\sigma\begin{bmatrix}0 & I\end{bmatrix}^T\end{bmatrix} = \begin{bmatrix}\mathcal{B}_{11}\\ 0\\ 0\end{bmatrix} \qquad (4.7.30)$$

with $\mathcal{B}_{11} = \mathcal{W}_1^T\Pi^T\mathcal{B}_r$. Furthermore, using (4.7.28) and the relations $\mathcal{A}_r(x)\mathcal{Y}_\nu = 0$ and

$$(I-\Pi_\infty)\mathcal{Y}_\sigma(\mathcal{Y}_\sigma^T\mathcal{A}_r\mathcal{Y}_\sigma)^{-\frac{1}{2}} = (I-\mathcal{Y}_\sigma(\mathcal{Y}_\sigma^T\mathcal{A}_r\mathcal{Y}_\sigma)^{-1}\mathcal{Y}_\sigma^T\mathcal{A}_r)\mathcal{Y}_\sigma(\mathcal{Y}_\sigma^T\mathcal{A}_r\mathcal{Y}_\sigma)^{-\frac{1}{2}} = 0,$$

we get

$$\begin{aligned}\mathcal{C}_r\mathcal{W} &= -\begin{bmatrix}0 & Z^T\end{bmatrix}\begin{bmatrix}\mathcal{W}_1^T\Pi^T(I-\Pi_\infty^T)\mathcal{A}_r(x)\\ (\mathcal{Y}_\nu^T\mathcal{E}_r\mathcal{Y}_\nu)^{-\frac{1}{2}}\mathcal{Y}_\nu^T\mathcal{A}_r(x)(I-\Pi_\infty)\\ (\mathcal{Y}_\sigma^T\mathcal{A}_r\mathcal{Y}_\sigma)^{-\frac{1}{2}}\mathcal{Y}_\sigma^T(I-\Pi_\infty^T)\mathcal{A}_r(x)\end{bmatrix}^T\\ &= \begin{bmatrix}\mathcal{C}_{11} & 0 & 0\end{bmatrix}\end{aligned} \qquad (4.7.31)$$

with

$$\mathcal{C}_{11} = -\begin{bmatrix}0 & Z^T\end{bmatrix}\mathcal{A}_r(x)(I-\Pi_\infty)\Pi\mathcal{W}_1 = -\begin{bmatrix}0 & Z^T\end{bmatrix}\mathcal{A}_r(I-\Pi_\infty)\mathcal{W}_1,$$

which is independent of the variable x. Introducing a new variable

$$\begin{bmatrix}x_1\\ x_2\\ x_3\end{bmatrix} = \mathcal{W}^{-1}x, \qquad (4.7.32)$$

we can rewrite (4.7.29) using (4.7.16), (4.7.30) and (4.7.31) as

$$\mathcal{E}_{11}\dot{x}_1 = \hat{\mathcal{A}}_{11}(x_1)x_1 + \mathcal{B}_{11}u, \tag{4.7.33a}$$
$$\dot{x}_2 = 0, \tag{4.7.33b}$$
$$0 = x_3, \tag{4.7.33c}$$
$$y = \mathcal{C}_{11}x_1, \tag{4.7.33d}$$

with $\hat{\mathcal{A}}_{11}(x_1) = \mathcal{A}_{11}(\Pi\mathcal{W}_1x_1 + \mathcal{Y}_\nu(\mathcal{Y}_\nu^T\mathcal{E}_r\mathcal{Y}_\nu)^{-\frac{1}{2}}x_2 + \mathcal{Y}_\sigma(\mathcal{Y}_\sigma^T\mathcal{A}_r\mathcal{Y}_\sigma)^{-\frac{1}{2}}x_3)$. Note that $x_3 = 0$ and x_2 is constant depending on the initial value. Therefore, $\hat{\mathcal{A}}_{11}$ dependents only on x_1. Since x_2 and x_3 do not contribute to the output, it is sufficient to consider equations (4.7.33a) and (4.7.33d), where the matrices $\mathcal{E}_{11}$ and $-\hat{\mathcal{A}}_{11}(x_1)$ are symmetric and positive definite. Thus, we have transformed the regular DAE system (4.7.9) into the ODE (4.7.33a), (4.7.33d).

In the context of the FIT discretization, the grad-div regularization of MQS systems has been considered in [Bos01, CSGB11, CW02] which is based on a spacial discretization of the Coulomb gauge (4.1.8). For other regularization techniques for MQS systems, we refer to [CM95, Hip00, Mun02].

4.7.3 Passivity

In this section, we examine the passivity of the weak formulation (4.7.1), the semidiscretized system (4.7.4), (4.7.5) and the regularized system (4.7.9), (4.7.10).

Theorem 4.25. *The variational MQS system* (4.7.1) *with the output* $y = \iota$ *is passive.*

Proof. The result can be proven analogously to Theorem 4.9. Consider a function $\vartheta : \Omega \times \mathbb{R}_0^+ \to \mathbb{R}_0^+$ given by

$$\vartheta(\xi, \varrho) = \frac{1}{2}\int_0^\varrho \boldsymbol{\nu}(\xi, \sqrt{\sigma})\, d\sigma = \int_0^{\sqrt{\varrho}} \boldsymbol{\nu}(\xi, \sigma)\sigma\, d\sigma \tag{4.7.34}$$

and define a storage function as

$$S(\mathbf{A}(\cdot, t)) = \int_\Omega \vartheta(\xi, \|\nabla \times \mathbf{A}(\xi, t)\|^2)\, d\xi.$$

This function is nonnegative, since $\boldsymbol{\nu}$ is positive. Furthermore, we have $S(0) = 0$. We now show $\frac{d}{dt}S(\mathbf{A}(\cdot, t)) \leqslant y(t)^Tv(t)$ for all v and suitable $\mathbf{A}$, $y = \iota$ that satisfy (4.7.1). We calculate

$$\begin{aligned}
\frac{d}{dt}S(\mathbf{A}(\cdot, t)) &= \frac{d}{dt}\int_\Omega \vartheta(\xi, \|\nabla \times \mathbf{A}(\xi, t)\|^2)\, d\xi \\
&= \int_\Omega \frac{\partial}{\partial\varrho}\vartheta(\xi, \|\nabla \times \mathbf{A}(\xi, t)\|^2)\frac{\partial}{\partial t}\|\nabla \times \mathbf{A}(\xi, t)\|^2\, d\xi \\
&= \int_\Omega \boldsymbol{\nu}(\xi, \|\nabla \times \mathbf{A}(\xi, t)\|)\Big(\nabla \times \mathbf{A}(\xi, t)\Big) \cdot \left(\frac{\partial}{\partial t}\nabla \times \mathbf{A}(\xi, t)\right) d\xi.
\end{aligned}$$

Taking $\mathbf{A}$ as a test function and $\frac{\partial}{\partial t}\mathbf{A}$ as a trial function, we obtain using equations (4.7.1) that

$$\begin{aligned}\frac{d}{dt}S(\mathbf{A}(\cdot,t)) &= -\int_\Omega \boldsymbol{\sigma}\frac{\partial}{\partial t}\mathbf{A}(\xi,t)\cdot\frac{\partial}{\partial t}\mathbf{A}(\xi,t)\,d\xi + \int_\Omega\left(\frac{\partial}{\partial t}\mathbf{A}(\xi,t)\right)^T \chi_{\mathrm{str}}\,d\xi\iota \\ &= -\int_\Omega \boldsymbol{\sigma}\frac{\partial}{\partial t}\mathbf{A}(\xi,t)\cdot\frac{\partial}{\partial t}\mathbf{A}(\xi,t)\,d\xi - \iota^T(t)\mathcal{R}\iota(t) + v^T(t)\iota(t) \\ &\leqslant y^T(t)v(t).\end{aligned}$$

Here, we used the property that the first two summands are nonpositive, since $\boldsymbol{\sigma}$ is nonnegative on Ω and $\mathcal{R}$ is positive definite. Integrating this inequality on $[0,T]$, we get the passivation inequality (2.3.3) and, hence, (4.7.1) is passive. □

The following theorem establishes that the spatial discretization of the variational MQS problem (4.7.1) preserves passivity.

Theorem 4.26. *The semidiscretized 3D MQS system* (4.7.4), (4.7.5) *is passive.*

Proof. The passivity of (4.7.4), (4.7.5) can be shown analogously to the proof of Theorem 4.10 by taking the storage function

$$S_d(a(t)) := \int_\Omega \vartheta(\xi, \|\nabla\times\sum_{i=1}^{n_e}\alpha_i(t)\psi_i^e(\xi)\|^2)\,d\xi$$

with ϑ as in (4.7.34). □

Finally, we show that the regularization of the semidiscretized system (4.7.4) preserves passivity.

Theorem 4.27. *The regularized 3D MQS system* (4.7.9), (4.7.10) *is passive.*

Proof. The result can be proved analogously to Theorem 4.10. Since a_{22} can be freely chosen and has no influence on the output, we take $a_{22}=0$. Then it yields

$$\begin{bmatrix} a_1 \\ a_2\end{bmatrix} = \mathcal{T}_4\begin{bmatrix} a_1 \\ a_{21}\end{bmatrix},$$

where

$$\mathcal{T}_4 = \begin{bmatrix} I & 0 \\ 0 & \hat{Y}_{\mathbf{C}_2}\end{bmatrix}.$$

Introducing new basis functions

$$[\phi_1(\xi),\ldots,\phi_{n_r}(\xi)] = [\psi_1^e(\xi),\ldots,\psi_{n_e}^e(\xi)]\mathcal{T}_4,$$

we obtain

$$\sum_{i=1}^{n_e}\alpha_i(t)\psi_i^e(\xi) = \sum_{i=1}^{n_r}\beta_i(t)\phi_i(\xi),$$

where

$$[\beta_1(t),\ldots,\beta_{n_1}(t)]^T = a_1(t) = [\alpha_1(t),\ldots,\alpha_{n_1}(t)]^T$$

and

$$[\beta_{n_1+1}(t), \dots, \beta_{n_r}(t)]^T = a_{21}(t).$$

For $x(t) = [\beta_1(t), \dots, \beta_{n_r}(t)]^T$, we define a storage function

$$S_r(x(t)) = \int_\Omega \vartheta(\xi, \|\nabla \times \sum_{i=1}^{n_r} \beta_i(t)\phi_i(\xi)\|^2)\, d\xi$$

with ϑ as in (4.7.34). This function is nonnegative, since $\boldsymbol{\nu}$ is positive and $S_1(0) = 0$ due to the definition of ϑ. We calculate

$$\begin{aligned}
&\frac{d}{dt} S_r(x(t)) = \frac{d}{dt} \int_\Omega \vartheta(\xi, \|\nabla \times \sum_{i=1}^{n_r} \beta_i(t)\phi_i(\xi)\|^2)\, d\xi \\
&= \int_\Omega \frac{\partial}{\partial \varrho} \vartheta(\xi, \|\nabla \times \sum_{i=1}^{n_r} \beta_i(t)\phi_i(\xi)\|^2) \frac{\partial}{\partial t} \|\nabla \times \sum_{i=1}^{n_r} \beta_i(t)\phi_i(\xi)\|^2 \, d\xi \\
&= \int_\Omega \boldsymbol{\nu}(\xi, \|\nabla \times \sum_{i=1}^{n_r} \beta_i(t)\phi_i(\xi)\|) \left(\nabla \times \sum_{i=1}^{n_r} \beta_i(t)\phi_i(\xi)\right) \cdot \left(\nabla \times \sum_{i=1}^{n_r} \dot\beta_i(t)\phi_i(\xi)\right) d\xi \\
&= \int_\Omega \boldsymbol{\nu}(\xi, \|\nabla \times \sum_{i=1}^{n_e} \alpha_i(t)\psi_i^e(\xi)\|) \left(\nabla \times \sum_{i=1}^{n_e} \alpha_i(t)\psi_i^e(\xi)\right) \cdot \left(\nabla \times \sum_{i=1}^{n_e} \dot\alpha_i(t)\psi_i^e(\xi)\right) d\xi \\
&= \dot{a}^T(t) \mathcal{K}(a(t)) a(t).
\end{aligned}$$

Using the relations

$$\begin{aligned}
\mathcal{A}_r(x) &= -\mathcal{T}_4^T \mathcal{K}(\mathcal{T}_4 x)\mathcal{T}_4, \\
\mathcal{E}_r &= \mathcal{T}_4^T(\mathcal{M} + X\mathcal{R}^{-1}X^T)\mathcal{T}_4 = \begin{bmatrix} \mathcal{M}_{11} & 0 \\ 0 & 0 \end{bmatrix} + \mathcal{B}_r \mathcal{R} \mathcal{B}_r^T,
\end{aligned}$$

and the equation (4.7.9a), we can continue

$$\begin{aligned}
\frac{d}{dt} S_r(x(t)) &= -\dot{x}^T(t)\mathcal{A}_r(x(t))x(t) \\
&= -\dot{x}^T(t)\mathcal{E}_r\dot{x}(t) + \dot{x}^T(t)\mathcal{B}_r u(t) \\
&= -\dot{a}_1^T(t)\mathcal{M}_{11}\dot{a}_1(t) - \dot{x}^T(t)\mathcal{B}_r\mathcal{R}\mathcal{B}_r^T\dot{x}(t) + \dot{x}^T(t)\mathcal{B}_r u(t) \\
&= -\dot{a}_1^T(t)\mathcal{M}_{11}\dot{a}_1(t) + \dot{x}^T(t)\mathcal{B}_r(u(t) - \mathcal{R}\mathcal{B}_r^T\dot{x}(t)).
\end{aligned}$$

Since the matrix $\mathcal{M}_{11}$ is positive definite, the first summand is negative. Furthermore, we use the output equation (4.7.9b) twice and obtain

$$\begin{aligned}
\frac{d}{dt} S_r(x(t)) &\leqslant (\mathcal{R}^{-1}u(t) - y(t))^T\mathcal{R}(\mathcal{R}^{-1}u(t) + y(t) - \mathcal{R}^{-1}u(t)) \\
&= -y^T(t)\mathcal{R}y(t) + y^T(t)u(t) \\
&\leqslant y^T(t)u(t).
\end{aligned}$$

Integrating this inequality on $[0, T]$, we get the passivation inequality

$$S_r(x(T)) - S_r(x(0)) \leqslant \int_0^T y^T(t)u(t)\, dt$$

which implies the passivity of the regularized system (4.7.9). □

It follows from Remark 2.17 and Theorems 4.26 and 4.27 that the semidiscretized MQS system (4.7.4) and its regularized form (4.7.9) are io-passive.

4.8 Model reduction for 3D linear MQS systems

Our goal is now to employ BT for model order reduction of the linear DAE system

$$\begin{aligned} \mathcal{E}_r \dot{x} &= \mathcal{A}_r x + \mathcal{B}_r u \\ y &= \mathcal{C}_r x \end{aligned} \tag{4.8.1}$$

with the system matrices

$$\begin{aligned} \mathcal{E}_r &= \mathcal{F}_\sigma \mathcal{M}_\sigma \mathcal{F}_\sigma^T, & \mathcal{A}_r &= -\mathcal{F}_\nu \mathcal{M}_\nu \mathcal{F}_\nu^T \\ \mathcal{B}_r &= \mathcal{F}_\nu \Upsilon \mathcal{R}^{-1}, & \mathcal{C}_r &= -\mathcal{B}_r \mathcal{E}_r^- \mathcal{A}_r. \end{aligned} \tag{4.8.2}$$

Unfortunately, we cannot use Algorithm 2.2 directly, because this system is stable but not asymptotically stable due to the fact that the pencil $\lambda\mathcal{E}_r - \mathcal{A}_r$ has zero eigenvalues. To overcome this difficulty, we proceed with the BT method developed in [RS11, Section 4].

First of all, note that system (4.8.1) is io-passive since its transfer function is positive real. The latter can be shown analogously to the 2D case. Using $\mathcal{C}_r = -\mathcal{B}_r^T \mathcal{E}_r^- \mathcal{A}_r$ and the properties of the pseudoinverse matrix $\mathcal{E}_r^-$ in (4.7.26), we can compute

$$\begin{aligned} \mathbf{G}(s) + \mathbf{G}^*(s) &= \mathcal{C}_r(s\mathcal{E}_r - \mathcal{A}_r)^{-1}\mathcal{B}_r + \mathcal{B}_r^T(\overline{s}\mathcal{E}_r - \mathcal{A}_r)^{-1}\mathcal{C}_r^T \\ &= -\mathcal{B}_r^T \mathcal{E}_r^- \mathcal{A}_r (s\mathcal{E}_r - \mathcal{A}_r)^{-1}\mathcal{B}_r - \mathcal{B}_r^T(\overline{s}\mathcal{E}_r - \mathcal{A}_r)^{-1}\mathcal{A}_r\mathcal{E}_r^-\mathcal{B}_r \\ &= F^*(s)(-(\overline{s}\mathcal{E}_r - \mathcal{A}_r)\mathcal{E}_r^-\mathcal{A}_r - \mathcal{A}_r\mathcal{E}_r^-(s\mathcal{E}_r - \mathcal{A}_r))F(s) \\ &= F^*(s)(-\overline{s}\mathcal{E}_r\mathcal{E}_r^-\mathcal{A}_r + \mathcal{A}_r\mathcal{E}_r^-\mathcal{A}_r - s\mathcal{A}_r\mathcal{E}_r^-\mathcal{E}_r + \mathcal{A}_r\mathcal{E}_r^-\mathcal{A}_r)F(s) \\ &= F^*(s)(2\mathcal{A}_r\mathcal{E}_r^-\mathcal{A}_r + 2\mathrm{Re}(s)\mathcal{E}_r\mathcal{E}_r^-(-\mathcal{A}_r)\mathcal{E}_r^-\mathcal{E}_r)F(s) \\ &\geqslant 0 \end{aligned}$$

for all $s \in \mathbb{C}^+$. Here, $F(s) = (s\mathcal{E}_r - \mathcal{A}_r)^{-1}\mathcal{B}_r$, and the matrices $\mathcal{A}_r\mathcal{E}_r^-\mathcal{A}_r$ and $\mathcal{E}_r\mathcal{E}_r^-(-\mathcal{A}_r)\mathcal{E}_r^-\mathcal{E}_r$ are both positive semidefinite. To apply BT, we first calculate the transfer function $\mathbf{G}(s)$ using the condensed form (4.7.16), which is the Weierstraß canonical form for the pencil $\lambda\mathcal{E}_r - \mathcal{A}_r$. Using (4.7.16), (4.7.30) and (4.7.31), we obtain

$$\begin{aligned} \mathbf{G}(s) &= \mathcal{C}_r(s\mathcal{E}_r - \mathcal{A}_r)^{-1}\mathcal{B}_r \\ &= \mathcal{C}_r\left(s\mathcal{W}^{-T}\begin{bmatrix} \mathcal{E}_{11} & & \\ & I & \\ & & 0 \end{bmatrix}\mathcal{W}^{-1} - \mathcal{W}^{-T}\begin{bmatrix} \mathcal{A}_{11} & & \\ & 0 & \\ & & I \end{bmatrix}\mathcal{W}^{-1}\right)^{-1}\mathcal{B}_r \\ &= \mathcal{C}_r\mathcal{W}\begin{bmatrix} s\mathcal{E}_{11} - \mathcal{A}_{11} & & \\ & sI & \\ & & -I \end{bmatrix}^{-1}\mathcal{W}^T\mathcal{B}_r \\ &= \mathcal{C}_{11}(s\mathcal{E}_{11} - \mathcal{A}_{11})^{-1}\mathcal{B}_{11}. \end{aligned}$$

This means that we need to reduce only the stable part

$$\begin{aligned} \mathcal{E}_{11}\dot{x}_1 &= \mathcal{A}_{11}x_1 + \mathcal{B}_{11}u, \\ y &= \mathcal{C}_{11}x_1, \end{aligned} \tag{4.8.3}$$

where $\mathcal{E}_{11}$ and $-\mathcal{A}_{11}$ are both symmetric and positive definite. The other parts can be removed from the system since they are uncontrollable and unobservable. System (4.8.3) is now asymptotically stable, and, hence, it can be reduced by the BT method as described in Section 2.4.1. Note, however, that the transformation matrix $\mathcal{W}$ is difficult to determine and the sparsity in $\mathcal{E}_{11}$ and $\mathcal{A}_{11}$ can not be expected anymore. Therefore, we never calculate the stable system (4.8.3) explicitly. Instead, we introduce the controllability and observability Gramians G_c and G_o for (4.8.1) as solutions of the projected continuous-time Lyapunov equations

$$\mathcal{E}_r G_c \mathcal{A}_r + \mathcal{A}_r G_c \mathcal{E}_r = -\Pi^T \mathcal{B}_r \mathcal{B}_r^T \Pi, \qquad G_c = \Pi G_c \Pi^T, \tag{4.8.4}$$

$$\mathcal{E}_r G_o \mathcal{A}_r + \mathcal{A}_r G_o \mathcal{E}_r = -\Pi^T \mathcal{C}_r^T \mathcal{C}_r \Pi, \qquad G_o = \Pi G_o \Pi^T, \tag{4.8.5}$$

where Π is the spectral projector onto the right deflating subspace of $\lambda\mathcal{E}_r - \mathcal{A}_r$ corresponding to the finite eigenvalues in the left complex half-plane. This projector is given in (4.7.19). Similarly to the 2D case, see Theorem 4.14, a relation between the controllability and the observability Gramians of system (4.8.1) can be established.

Theorem 4.28. *Consider the DAE system* (4.8.1), (4.8.2). *Let G_c and G_o be the controllability and observability Gramians of* (4.8.1) *which solve the projected continuous-time Lyapunov equations* (4.8.4) *and* (4.8.5). *Then*

$$\mathcal{E}_r G_o \mathcal{E}_r = \mathcal{A}_r G_c \mathcal{A}_r.$$

Proof. Consider the reflexive inverse $\mathcal{E}_r^-$ of $\mathcal{E}_r$ given in (4.7.20) and the reflexive inverse of $\mathcal{A}_r$ given by

$$\mathcal{A}_r^- = \mathcal{W} \begin{bmatrix} \mathcal{A}_{11}^{-1} & & \\ & 0 & \\ & & I \end{bmatrix} \mathcal{W}^T.$$

Then multiplying the Lyapunov equation (4.8.4) (resp. (4.8.5)) from the left and right with $\mathcal{E}_r^-$ (resp. with $\mathcal{A}_r^-$) and using the relations (4.7.24), (4.7.25) and

$$(I-\Pi_\infty)\Pi = \Pi, \qquad \mathcal{A}_r^-\mathcal{A}_r = I - \Pi_0, \qquad (I-\Pi_0)\Pi = \Pi,$$
$$\Pi\mathcal{E}_r^- = \mathcal{E}_r^-\Pi^T, \qquad \Pi\mathcal{A}_r^- = \mathcal{A}_r^-\Pi^T,$$

we obtain

$$\mathcal{A}_r^-(\mathcal{A}_r G_c \mathcal{A}_r)\mathcal{E}_r^- + \mathcal{E}_r^-(\mathcal{A}_r G_c \mathcal{A}_r)\mathcal{A}_r^- = -\Pi\mathcal{E}_r^-\mathcal{B}_r\mathcal{B}_r^T\mathcal{E}_r^-\Pi^T, \quad G_c = \Pi G_c \Pi^T, \tag{4.8.6}$$

$$\mathcal{A}_r^-(\mathcal{E}_r G_o \mathcal{E}_r)\mathcal{E}_r^- + \mathcal{E}_r^-(\mathcal{E}_r G_o \mathcal{E}_r)\mathcal{A}_r^- = -\mathcal{A}_r^-\Pi^T\mathcal{C}_r^T\mathcal{C}_r\Pi\mathcal{A}_r^-, \quad G_o = \Pi G_o \Pi^T. \tag{4.8.7}$$

Furthermore, it follows from (4.7.28) and $\Pi\mathcal{E}_r^-\mathcal{A}_r = \mathcal{E}_r^-\mathcal{A}_r\Pi$ that

$$\begin{aligned}\mathcal{C}_r\Pi\mathcal{A}_r^- &= -\mathcal{B}_r^T\mathcal{E}_r^-\mathcal{A}_r\Pi\mathcal{A}_r^- = -\mathcal{B}_r^T\Pi\mathcal{E}_r^-\mathcal{A}_r\mathcal{A}_r^- \\ &= -\mathcal{B}_r^T\Pi\mathcal{E}_r^-(I-\Pi_0^T) = -\mathcal{B}_r^T\Pi\mathcal{E}_r^- = -\mathcal{B}_r^T\mathcal{E}_r^-\Pi^T.\end{aligned}$$

Then equation (4.8.7) can be written as

$$\mathcal{A}_r^-(\mathcal{E}_r G_o \mathcal{E}_r)\mathcal{E}_r^- + \mathcal{E}_r^-(\mathcal{E}_r G_o \mathcal{E}_r)\mathcal{A}_r^- = -\Pi\mathcal{E}_r^-\mathcal{B}_r\mathcal{B}_r^T\mathcal{E}_r^-\Pi^T, \qquad G_o = \Pi G_o \Pi^T. \tag{4.8.8}$$

Since $\mathcal{E}_r^-$ and $-\mathcal{A}_r^-$ are symmetric and positive definite and Π^T is the spectral projector onto the right deflating subspace of $\lambda\mathcal{E}_r^- - \mathcal{A}_r^-$ corresponding to the eigenvalues in the left half-plane, the Lyapunov equations (4.8.6) and (4.8.8) are uniquely solvable, and, hence, $\mathcal{E}_r G_o \mathcal{E}_r = \mathcal{A}_r G_c \mathcal{A}_r$. □

Theorem 4.28 implies that we need to solve only the projected Lyapunov equation (4.8.4) for the Cholesky factor of $G_c = Z_c Z_c^T$. Then it follows from the relation

$$G_o = \mathcal{E}_r^- \mathcal{A}_r G_c \mathcal{A}_r \mathcal{E}_r^- = (-\mathcal{E}_r^- \mathcal{A}_r Z_c)(-Z_c^T \mathcal{A}_r \mathcal{E}_r^-)$$

that the Cholesky factor Z_o of the observability Gramian G_o can be calculated as $Z_o = -\mathcal{E}_r^- \mathcal{A}_r Z_c$. For the Cholesky factor Z_c it holds that

$$\mathcal{E}_r^- \mathcal{E}_r Z_c = (I - \Pi_\infty) Z_c = (I - \Pi_\infty)\Pi Z_c = \Pi Z_c = Z_c.$$

Therefore, the Hankel singular values can be computed from the EVD

$$Z_o^T \mathcal{E}_r Z_c = (-Z_c^T \mathcal{A}_r \mathcal{E}_r^-)\mathcal{E}_r Z_c = -Z_c^T \mathcal{A}_r Z_c = \begin{bmatrix} U_1 & U_2 \end{bmatrix} \begin{bmatrix} \Lambda_1 & \\ & \Lambda_2 \end{bmatrix} \begin{bmatrix} U_1 & U_2 \end{bmatrix}^T.$$

Then, similarly to the 2D case, the reduced-order model

$$\begin{aligned} \tilde{\mathcal{E}}_r \dot{\tilde{x}} &= \tilde{\mathcal{A}}_r \tilde{x} + \tilde{\mathcal{B}}_r u, \\ \tilde{y} &= \tilde{\mathcal{C}}_r \tilde{x}, \end{aligned} \tag{4.8.9}$$

can be computed by projection

$$\tilde{\mathcal{E}}_r = W^T \mathcal{E}_r V, \qquad \tilde{\mathcal{A}}_r = W^T \mathcal{A}_r V, \qquad \tilde{\mathcal{B}}_r = W^T \mathcal{B}_r, \qquad \tilde{\mathcal{C}}_r = \mathcal{C}_r V$$

with the projection matrices $V = Z_c U_1 \Lambda_1^{-\frac{1}{2}}$ and $W = Z_o U_1 \Lambda_1^{-\frac{1}{2}} = -\mathcal{E}_r^- \mathcal{A}_r V$. The reduced matrices have the form

$$\begin{aligned} \tilde{\mathcal{E}}_r &= -V^T \mathcal{A}_r \mathcal{E}_r^- \mathcal{E}_r V = -\Lambda_1^{-\frac{1}{2}} U_1^T Z_c^T \mathcal{A}_r Z_c U_1 \Lambda_1^{-\frac{1}{2}} = I, \\ \tilde{\mathcal{A}}_r &= -V^T \mathcal{A}_r \mathcal{E}_r^- \mathcal{A}_r V, \\ \tilde{\mathcal{B}}_r &= -V^T \mathcal{A}_r \mathcal{E}_r^- \mathcal{B}_r = V^T \mathcal{C}_r^T = \tilde{\mathcal{C}}_r^T. \end{aligned}$$

Since $\mathcal{E}_r$ and $-\mathcal{A}_r$ are symmetric and positive definite, the matrices $\tilde{\mathcal{E}}_r$ and $-\tilde{\mathcal{A}}_r$ are symmetric and positive definite too. Then it follows from Theorem 2.15 that the reduced system (4.8.9) is io-passive. The BT method for the DAE system (4.8.1) is presented in Algorithm 4.5. Note that in the LR-ADI method used in Step 3 for solving the projected Lyapunov equation (4.8.4), we need to compute the projector-vector products with Π_0 and Π_∞. We will first discuss the computation of the product $\mathcal{Y}_\sigma(\mathcal{Y}_\sigma^T \mathcal{A}_r \mathcal{Y}_\sigma)^{-1}\mathcal{Y}_\sigma^T w$ for a vector w.

Lemma 4.29. *Let $\mathcal{A}_r$ be given as in (4.7.10) with a constant matrix $\mathcal{K}_{11}$, $\mathcal{Y}_\sigma$ a basis of $\ker(\mathcal{F}_\sigma^T)$ and $w = \begin{bmatrix} w_1^T, & w_2^T \end{bmatrix}^T \in \mathbb{R}^{n_r}$. Then the vector $\hat{z} = \mathcal{Y}_\sigma(\mathcal{Y}_\sigma^T \mathcal{A}_r \mathcal{Y}_\sigma)^{-1}\mathcal{Y}_\sigma^T w$ can be determined as $\hat{z} = \begin{bmatrix} 0, & z_1^T \end{bmatrix}^T$, where $\begin{bmatrix} z_1^T, & z_2^T \end{bmatrix}^T$ satisfies the linear system*

$$\begin{bmatrix} -\hat{Y}_{\mathbf{C}_2}^T \mathcal{K}_{22} \hat{Y}_{\mathbf{C}_2} & \hat{Y}_{\mathbf{C}_2}^T \mathcal{X}_2 \\ \mathcal{X}_2^T \hat{Y}_{\mathbf{C}_2} & 0 \end{bmatrix} \begin{bmatrix} z_1 \\ z_2 \end{bmatrix} = \begin{bmatrix} w_2 \\ 0 \end{bmatrix}. \tag{4.8.10}$$

Algorithm 4.5: Balanced truncation for the 3D linear MQS system

Input : $\mathcal{M}_{11}, \mathcal{K}_{11} \in \mathbb{R}^{n_1 \times n_1}$, $\mathcal{K}_{12} \in \mathbb{R}^{n_1 \times n_2}$, $\mathcal{K}_{21} \in \mathbb{R}^{n_2 \times n_1}$, $\mathcal{K}_{22} \in \mathbb{R}^{n_2 \times n_2}$, $\mathcal{X}_1 \in \mathbb{R}^{n_1 \times m}$, $\mathcal{X}_2 \in \mathbb{R}^{n_2 \times m}$, $\mathcal{R} \in \mathbb{R}^{m \times m}$.

Output: a reduced-order asymptotically stable system $(\tilde{\mathcal{E}}_r, \tilde{\mathcal{A}}_r, \tilde{\mathcal{B}}_r, \tilde{\mathcal{C}}_r)$.

1 Compute a basis matrix $\hat{Y}_{\mathbf{C}_2}$ as discussed in Remark 4.22

2 Compute $\mathcal{E}_r$, $\mathcal{A}_r$, $\mathcal{B}_r$ given in (4.7.10).

3 Solve the projected Lyapunov equation (4.8.4) for the Cholesky factor $\tilde{Z}_c$ of $G_c \approx \tilde{Z}_c \tilde{Z}_c^T$ using the LR-ADI method as described in [Sty08].

4 Compute the EVD

$$-\tilde{Z}_c^T \mathcal{A}_r \tilde{Z}_c = \begin{bmatrix} U_1 & U_2 \end{bmatrix} \begin{bmatrix} \Lambda_1 & 0 \\ 0 & \Lambda_2 \end{bmatrix} \begin{bmatrix} U_1 & U_2 \end{bmatrix}^T.$$

5 Compute the projection matrix $V = \tilde{Z}_c U_1 \Lambda_1^{-\frac{1}{2}}$.

6 Compute the reduced matrices $\tilde{\mathcal{E}}_r = I$, $\tilde{\mathcal{A}}_r = -V^T \mathcal{A}_r \mathcal{E}_r^- \mathcal{A}_r V$, $\tilde{\mathcal{B}}_r = -V^T \mathcal{A}_r \mathcal{E}_r^- \mathcal{B}_r$ and $\tilde{\mathcal{C}}_r = \tilde{\mathcal{B}}_r^T$.

Proof. We first show that $\hat{z} = \mathcal{Y}_\sigma (\mathcal{Y}_\sigma^T \mathcal{A}_r \mathcal{Y}_\sigma)^{-1} \mathcal{Y}_\sigma^T w$ if and only if

$$\begin{bmatrix} \mathcal{A}_r & \hat{\mathcal{Y}}_\sigma \\ \hat{\mathcal{Y}}_\sigma^T & 0 \end{bmatrix} \begin{bmatrix} \hat{z} \\ \hat{z}_2 \end{bmatrix} = \begin{bmatrix} w \\ 0 \end{bmatrix}, \tag{4.8.11}$$

where the columns of $\hat{\mathcal{Y}}_\sigma$ form a basis of $\operatorname{im}(\mathcal{F}_\sigma)$. Assume that $\begin{bmatrix} \hat{z}^T & \hat{z}_2^T \end{bmatrix}^T$ solves (4.8.11). Then $\hat{\mathcal{Y}}_\sigma^T \hat{z} = 0$ and, therefore, $\hat{z} \in \ker(\hat{\mathcal{Y}}_\sigma^T) = \operatorname{im}(\mathcal{Y}_\sigma)$. This means that there exists $\tilde{z}$ such that $\hat{z} = \mathcal{Y}_\sigma \tilde{z}$. Inserting this vector into the first equation in (4.8.11), we obtain

$$\mathcal{A}_r \mathcal{Y}_\sigma \tilde{z} + \hat{\mathcal{Y}}_\sigma \hat{z}_2 = w.$$

Multiplying this equation from the left with $\mathcal{Y}_\sigma^T$, we find

$$\tilde{z} = (\mathcal{Y}_\sigma^T \mathcal{A}_r \mathcal{Y}_\sigma)^{-1} \mathcal{Y}_\sigma^T w$$

and, therefore, $\hat{z} = \mathcal{Y}_\sigma \tilde{z} = \mathcal{Y}_\sigma (\mathcal{Y}_\sigma^T \mathcal{A}_r \mathcal{Y}_\sigma)^{-1} \mathcal{Y}_\sigma^T w$.

We now show the opposite direction. For $\hat{z} = \mathcal{Y}_\sigma (\mathcal{Y}_\sigma^T \mathcal{A}_r \mathcal{Y}_\sigma)^{-1} \mathcal{Y}_\sigma^T w$ and

$$\hat{z}_2 = (\hat{\mathcal{Y}}_\sigma^T \hat{\mathcal{Y}}_\sigma)^{-1} \hat{\mathcal{Y}}_\sigma^T (w - \mathcal{A}_r \hat{z}),$$

we have $\hat{\mathcal{Y}}_\sigma \hat{z} = \hat{\mathcal{Y}}_\sigma^T \mathcal{Y}_\sigma (\mathcal{Y}_\sigma^T \mathcal{A}_r \mathcal{Y}_\sigma)^{-1} \mathcal{Y}_\sigma^T w = 0$ since $\hat{\mathcal{Y}}_\sigma^T \mathcal{Y}_\sigma = 0$ and

$$\begin{aligned} \mathcal{A}_r \hat{z} + \hat{\mathcal{Y}}_\sigma^T \hat{z}_2 &= \mathcal{A}_r \hat{z} + \hat{\mathcal{Y}}_\sigma^T (\hat{\mathcal{Y}}_\sigma^T \hat{\mathcal{Y}}_\sigma)^{-1} \hat{\mathcal{Y}}_\sigma^T (w - \mathcal{A}_r \hat{z}) \\ &= (I - \hat{\mathcal{Y}}_\sigma (\hat{\mathcal{Y}}_\sigma^T \hat{\mathcal{Y}}_\sigma)^{-1} \hat{\mathcal{Y}}_\sigma^T) \mathcal{A}_r \hat{z} + \hat{\mathcal{Y}}_\sigma (\hat{\mathcal{Y}}_\sigma^T \hat{\mathcal{Y}}_\sigma)^{-1} \hat{\mathcal{Y}}_\sigma^T w. \end{aligned}$$

Using $\hat{\mathcal{Y}}_\sigma (\hat{\mathcal{Y}}_\sigma^T \hat{\mathcal{Y}}_\sigma)^{-1} \hat{\mathcal{Y}}_\sigma^T + \mathcal{Y}_\sigma (\mathcal{Y}_\sigma^T \mathcal{Y}_\sigma)^{-1} \mathcal{Y}_\sigma^T = I$, we obtain

$$\begin{aligned} \mathcal{A}_r \hat{z} + \hat{\mathcal{Y}}_\sigma \hat{z}_2 &= \mathcal{Y}_\sigma (\mathcal{Y}_\sigma^T \mathcal{Y}_\sigma)^{-1} \mathcal{Y}_\sigma^T \mathcal{A}_r \hat{z} + \hat{\mathcal{Y}}_\sigma (\hat{\mathcal{Y}}_\sigma^T \hat{\mathcal{Y}}_\sigma)^{-1} \hat{\mathcal{Y}}_\sigma^T w \\ &= \mathcal{Y}_\sigma (\mathcal{Y}_\sigma^T \mathcal{Y}_\sigma)^{-1} \mathcal{Y}_\sigma^T \mathcal{A}_r \mathcal{Y}_\sigma (\mathcal{Y}_\sigma^T \mathcal{A}_r \mathcal{Y}_\sigma)^{-1} \mathcal{Y}_\sigma^T w + \hat{\mathcal{Y}}_\sigma (\hat{\mathcal{Y}}_\sigma^T \hat{\mathcal{Y}}_\sigma)^{-1} \hat{\mathcal{Y}}_\sigma^T w \\ &= w \end{aligned}$$

and, therefore, $\left[\hat{z}^T,\, \hat{z}_2^T\right]^T$ solves equation (4.8.11).

As it has already been shown in the proof of Theorem 4.23, the matrix $\hat{Y}_{\mathbf{C}_2}^T \mathcal{X}_2$ has full column rank. Then it follows from

$$\mathcal{F}_\sigma = \begin{bmatrix} I & 0 \\ 0 & \hat{Y}_{\mathbf{C}_2}^T \mathcal{X}_2 \end{bmatrix} \begin{bmatrix} I & \mathcal{X}_1 \\ 0 & I \end{bmatrix}$$

that the columns of $\hat{\mathcal{Y}}_\sigma = \begin{bmatrix} I & 0 \\ 0 & \hat{Y}_{\mathbf{C}_2}^T \mathcal{X}_2 \end{bmatrix}$ form a basis of $\operatorname{im}(\mathcal{F}_\sigma)$. With this structure, equation (4.8.11) yields

$$\begin{bmatrix} -\mathcal{K}_{11} & -\mathcal{K}_{12}\hat{Y}_{\mathbf{C}_2} & I & 0 \\ -\hat{Y}_{\mathbf{C}_2}^T \mathcal{K}_{21} & -\hat{Y}_{\mathbf{C}_2}^T \mathcal{K}_{22}\hat{Y}_{\mathbf{C}_2} & 0 & \hat{Y}_{\mathbf{C}_2}^T \mathcal{X}_2 \\ I & 0 & 0 & 0 \\ 0 & \mathcal{X}_2^T \hat{Y}_{\mathbf{C}_2} & 0 & 0 \end{bmatrix} \begin{bmatrix} \bar{z}_1 \\ z_1 \\ \bar{z}_2 \\ z_2 \end{bmatrix} = \begin{bmatrix} w_1 \\ w_2 \\ 0 \\ 0 \end{bmatrix}, \tag{4.8.12}$$

with $\hat{z} = \left[\bar{z}_1^T,\, z_1^T\right]^T$, $\hat{z}_2 = \left[\bar{z}_2^T,\, z_2^T\right]^T$ and $w = \left[w_1^T,\, w_2^T\right]^T$. Therefore, $\bar{z}_1 = 0$ is valid, and we only have to determine z_1. Thus, equation (4.8.12) can be reduced to equation (4.8.10) whose solution gives $\hat{z} = \left[0,\, z_1^T\right]^T$. Note that this equation is uniquely solvable, since (4.8.11) is uniquely solvable. □

The product $\Pi_\infty w_1 = \mathcal{Y}_\sigma(\mathcal{Y}_\sigma^T \mathcal{A}_r \mathcal{Y}_\sigma)^{-1}\mathcal{Y}_\sigma^T \mathcal{A}_r w_1$ can be determined by applying Lemma 4.29 with $w = \mathcal{A}_r w_1$, and $\Pi_\infty^T w$ can be computed as $\Pi_\infty^T w = \mathcal{A}_r \left[0,\, z_1^T\right]^T$, where z_1 is defined as in Lemma 4.29.

Similarly to the proof of Lemma 4.29, one can show that the vector $\hat{z} = \mathcal{Y}_\nu(\mathcal{Y}_\nu^T \mathcal{E}_r \mathcal{Y}_\nu)^{-1}\mathcal{Y}_\nu^T w$ can be determined by solving the linear system

$$\begin{bmatrix} \mathcal{E}_r & \hat{\mathcal{Y}}_\nu \\ \hat{\mathcal{Y}}_\nu^T & 0 \end{bmatrix} \begin{bmatrix} \hat{z} \\ \hat{z}_2 \end{bmatrix} = \begin{bmatrix} w \\ 0 \end{bmatrix}, \tag{4.8.13}$$

where the columns of $\hat{\mathcal{Y}}_\nu$ form a basis of $\operatorname{im}(\mathcal{F}_\nu)$. This basis matrix can be computed from a sparse LU decomposition of $\mathcal{F}_\nu$ as proposed in [Kow06]. Then the product $\Pi_0 w_1 = \mathcal{Y}_\nu(\mathcal{Y}_\nu^T \mathcal{E}_r \mathcal{Y}_\nu)^{-1}\mathcal{Y}_\nu^T \mathcal{E}_r w_1$ can be calculated by solving (4.8.13) with $w = \mathcal{E}_r w_1$. To determine $\Pi_0^T w$, we have to solve (4.8.13) for $\hat{z}$ and compute $\Pi_0^T w = \mathcal{E}_r \hat{z}$.

In Step 4 of Algorithm 4.5, we also need to compute the product $\mathcal{E}_r^- \mathcal{A}_r w$. The following lemma presents how to calculate such a product in an efficient way.

Lemma 4.30. *Let $\mathcal{E}_r$ and $\mathcal{A}_r$ be given as in (4.7.10), $Z = \hat{Y}_{\mathbf{C}_2}^T \mathcal{X}_2(\mathcal{X}_2^T \hat{Y}_{\mathbf{C}_2} \hat{Y}_{\mathbf{C}_2}^T \mathcal{X}_2)^{-1}$ and $w = \left[w_1^T,\, w_2^T\right]^T \in \mathbb{R}^{n_r}$. Then the vector $z = \left[z_1^T,\, z_2^T\right]^T = \mathcal{E}_r^- w$ can be determined by solving the linear system*

$$\begin{bmatrix} \mathcal{M}_{11} + \mathcal{X}_1 \mathcal{R}^{-1}\mathcal{X}_1^T & \mathcal{X}_1 \mathcal{R}^{-1}\mathcal{X}_2^T \hat{Y}_{\mathbf{C}_2} & 0 \\ \mathcal{R}^{-1}\mathcal{X}_1^T & \mathcal{R}^{-1}\mathcal{X}_2^T \hat{Y}_{\mathbf{C}_2} & 0 \\ \hat{Y}_{\mathbf{C}_2}^T \mathcal{K}_{21} & \hat{Y}_{\mathbf{C}_2}^T \mathcal{K}_{22}\hat{Y}_{\mathbf{C}_2} & \hat{Y}_{\mathbf{C}_2}^T \mathcal{X}_2 \end{bmatrix} \begin{bmatrix} z_1 \\ z_2 \\ z_3 \end{bmatrix} = \begin{bmatrix} w_1 - \mathcal{K}_{12}\hat{Y}_{\mathbf{C}_2}\hat{w}_1 \\ Z^T w_2 - Z^T \hat{Y}_{\mathbf{C}_2}^T \mathcal{K}_{22}\hat{Y}_{\mathbf{C}_2}\hat{w}_1 \\ 0 \end{bmatrix} \tag{4.8.14}$$

with $\hat{w} = \begin{bmatrix}\hat{w}_1^T & \hat{w}_2^T\end{bmatrix}^T$ *satisfying*

$$\begin{bmatrix} -\hat{Y}_{\mathbf{C}_2}^T \mathcal{K}_{22} \hat{Y}_{\mathbf{C}_2} & \hat{Y}_{\mathbf{C}_2}^T \mathcal{X}_2 \\ \mathcal{X}_2^T \hat{Y}_{\mathbf{C}_2} & 0 \end{bmatrix} \begin{bmatrix} \hat{w}_1 \\ \hat{w}_2 \end{bmatrix} = \begin{bmatrix} w_2 \\ 0 \end{bmatrix}. \tag{4.8.15}$$

Proof. Let $\left[z_1^T, z_2^T, z_3^T\right]^T$ and $\left[\hat{w}_1^T, \hat{w}_2^T\right]^T$ be given such that equations (4.8.14) and (4.8.15) are fulfilled. Applying Lemma 4.29 to equation (4.8.15), we obtain

$$\mathcal{Y}_\sigma(\mathcal{Y}_\sigma^T \mathcal{A}_r \mathcal{Y}_\sigma)^{-1}\mathcal{Y}_\sigma^T w = \begin{bmatrix} 0 \\ \hat{w}_1 \end{bmatrix}.$$

Therefore, the first two lines of (4.8.14) are equivalent to

$$\begin{bmatrix} I & 0 \\ 0 & Z^T \end{bmatrix} \mathcal{E}_r z = \begin{bmatrix} I & 0 \\ 0 & Z^T \end{bmatrix} (I - \Pi_\infty^T) w.$$

Furthermore, it holds that $\mathcal{Y}_\sigma^T \mathcal{E}_r z = 0 = \mathcal{Y}_\sigma^T (I - \Pi_\infty^T) w$ and with the columns of $\mathcal{Y}_\sigma$ forming a basis of $\ker(\mathcal{F}_\sigma)$ we have $\left[\begin{bmatrix} I & 0 \\ 0 & Z \end{bmatrix} \ \mathcal{Y}_\sigma\right]$ invertible as $\begin{bmatrix} I & 0 \\ 0 & Z \end{bmatrix}$ is a basis of $\operatorname{im}(\mathcal{F}_\sigma)$. To sum up, we have $\mathcal{E}_r z = (I - \Pi_\infty^T) w$. Multiplying this equation from left with $\mathcal{E}_r^-$ and using equations (4.7.25), (4.7.24) and (4.7.23) we get

$$(I - \Pi_\infty) z = \mathcal{E}_r^- \mathcal{E}_r z = \mathcal{E}_r^- (I - \Pi_\infty^T) w = \mathcal{E}_r^- \mathcal{E}_r \mathcal{E}_r^- v = \mathcal{E}_r^- w. \tag{4.8.16}$$

Therefore, it remains to show that $(I - \Pi_\infty) z = z$. Applying Lemma 4.29 to compute $\Pi_\infty z$, we have $\Pi_\infty z = \left[0, \, z_4^T\right]^T$ with $\left[z_4^T, \, z_3^T\right]^T$ solving

$$\begin{bmatrix} -\hat{Y}_{\mathbf{C}_2}^T \mathcal{K}_{22} \hat{Y}_{\mathbf{C}_2} & \hat{Y}_{\mathbf{C}_2}^T \mathcal{X}_2 \\ \mathcal{X}_2^T \hat{Y}_{\mathbf{C}_2} & 0 \end{bmatrix} \begin{bmatrix} z_4 \\ z_3 \end{bmatrix} = \begin{bmatrix} -\hat{Y}_{\mathbf{C}_2}^T \mathcal{K}_{21} z_1 - \hat{Y}_{\mathbf{C}_2}^T \mathcal{K}_{22} \hat{Y}_{\mathbf{C}_2} z_2 \\ 0 \end{bmatrix}.$$

As the third line of (4.8.14) holds, we have $z_4 = 0$ since

$$\hat{Y}_{\mathbf{C}_2}^T \mathcal{K}_{22} \hat{Y}_{\mathbf{C}_2} = \hat{Y}_{\mathbf{C}_2}^T \mathbf{C}_2^T \mathcal{M}_\nu \mathbf{C}_2 \hat{Y}_{\mathbf{C}_2}$$

is nonsingular. Therefore, $(I - \Pi_\infty) z = z$. Thus, from (4.8.16) it follows that $z = \mathcal{E}_r^- w$. □

Finally, we discuss the computation of $z = (\tau \mathcal{E}_r + \mathcal{A}_r)^{-1} w$ required in the LR-ADI method in Step 3 of Algorithm 4.5. If $\tau \mathcal{E}_r + \mathcal{A}_r$ remains sparse, we just solve the linear system $(\tau \mathcal{E}_r + \mathcal{A}_r) z = w$. If $\tau \mathcal{E}_r + \mathcal{A}_r$ gets fill-in due to the multiplication with $\hat{Y}_{\mathbf{C}_2}$, then we can use the following lemma to compute $z = (\tau \mathcal{E}_r + \mathcal{A}_r)^{-1} w$.

Lemma 4.31. *Let* $\mathcal{E}_r$ *and* $\mathcal{A}_r$ *be as in* (4.7.7) *with a constant matrix* $\mathcal{K}_{11}$ *and* $\tau \in \mathbb{C}^-$. *Then the vector* $z = (\tau \mathcal{E}_r + \mathcal{A}_r)^{-1} \left[w_1^T, \, w_2^T\right]$ *can be determined as*

$$z = \left[z_1^T, \left((\hat{Y}_{\mathbf{C}_2}^T \hat{Y}_{\mathbf{C}_2})^{-1} \hat{Y}_{\mathbf{C}_2}^T z_2\right)^T\right]^T,$$

where z_1 *and* z_2 *satisfy the linear system*

$$\begin{bmatrix} \tau \mathcal{M}_{11} - \mathcal{K}_{11} & -\mathcal{K}_{12} & \mathcal{X}_1 & 0 \\ -\mathcal{K}_{21} & -\mathcal{K}_{22} & \mathcal{X}_2 & Y_{\mathbf{C}_2} \\ \tau \mathcal{X}_1^T & \tau \mathcal{X}_2^T & -\mathcal{R} & 0 \\ 0 & Y_{\mathbf{C}_2}^T & 0 & 0 \end{bmatrix} \begin{bmatrix} z_1 \\ z_2 \\ z_3 \\ z_4 \end{bmatrix} = \begin{bmatrix} w_1 \\ \hat{Y}_{\mathbf{C}_2} (\hat{Y}_{\mathbf{C}_2}^T \hat{Y}_{\mathbf{C}_2})^{-1} w_2 \\ 0 \\ 0 \end{bmatrix} \tag{4.8.17}$$

with the basis matrices $Y_{\mathbf{C}_2}$ *and* $\hat{Y}_{\mathbf{C}_2}$ *of* $\ker(\mathbf{C}_2)$ *and* $\operatorname{im}(\mathbf{C}_2^T)$, *respectively.*

Proof. First, note that due to the choice of $Y_{\mathbf{C}_2}$ the coefficient matrix in system (4.8.17) is nonsingular. This system can be written as

$$
\begin{aligned}
(\tau\mathcal{M}_{11} - \mathcal{K}_{11})z_1 - \mathcal{K}_{12}z_2 + \mathcal{X}_1 z_3 \qquad &= w_1, && (4.8.18a)\\
-\mathcal{K}_{21}z_1 - \mathcal{K}_{22}z_2 + \mathcal{X}_2 z_3 + Y_{\mathbf{C}_2} z_4 &= \hat{Y}_{\mathbf{C}_2}(\hat{Y}_{\mathbf{C}_2}^T \hat{Y}_{\mathbf{C}_2})^{-1} w_2, && (4.8.18b)\\
\tau\mathcal{X}_1^T z_1 + \tau\mathcal{X}_2^T z_2 - \mathcal{R} z_3 \qquad &= 0, && (4.8.18c)\\
Y_{\mathbf{C}_2}^T z_2 \qquad &= 0. && (4.8.18d)
\end{aligned}
$$

It follows from (4.8.18d) that $z_2 \in \ker(Y_{\mathbf{C}_2}^T) = \operatorname{im}(\hat{Y}_{\mathbf{C}_2})$. Then there exists $\hat{z}_2$ such that $z_2 = \hat{Y}_{\mathbf{C}_2}\hat{z}_2$. This relation implies

$$
\hat{z}_2 = (\hat{Y}_{\mathbf{C}_2}^T \hat{Y}_{\mathbf{C}_2})^{-1} \hat{Y}_{\mathbf{C}_2}^T z_2. \tag{4.8.19}
$$

Further, from equation (4.8.18c) we have $z_3 = \tau\mathcal{R}^{-1}\mathcal{X}_1^T z_1 + \tau\mathcal{R}^{-1}\mathcal{X}_2^T z_2$. Substituting z_2 and z_3 into (4.8.18a) and (4.8.18b) and multiplying equation (4.8.18b) from the left with $\hat{Y}_{\mathbf{C}_2}^T$, we obtain

$$
(\tau\mathcal{E}_r + \mathcal{A}_r)\begin{bmatrix} z_1 \\ \hat{z}_2 \end{bmatrix} = \begin{bmatrix} w_1 \\ w_2 \end{bmatrix}.
$$

This equation together with (4.8.19) implies that

$$
\begin{bmatrix} z_1 \\ (\hat{Y}_{\mathbf{C}_2}^T \hat{Y}_{\mathbf{C}_2})^{-1} \hat{Y}_{\mathbf{C}_2}^T z_2 \end{bmatrix} = (\tau\mathcal{E}_r + \mathcal{A}_r)^{-1} \begin{bmatrix} w_1 \\ w_2 \end{bmatrix}.
$$

□

Remark 4.32. We have seen that BT for the 3D linear MQS problem (4.7.4) is applicable analogously to BT for the 2D linear MQS problem (4.3.5). Model order reduction using POD and DEIM for the nonlinear MQS systems can also be easily transferred from the 2D system (4.3.5) to the regularized 3D system (4.7.9).

5 Summary and outlook

The simulation of modern electronic devices is becoming increasingly important in their development. At the same time, models are becoming more accurate and complex. They are often available as coupled systems of partial differential-algebraic equations. Model order reduction and dynamic iteration are two tools for the efficient numerical solution of such systems. In the first part of this thesis, we studied the DIRM method for coupled systems of nonlinear ordinary differential equations (ODEs), which is based on a combination of the dynamic iteration and model reduction methods. It was shown that the DIRM procedure heavily relies on an appropriate strategy for the choice of method parameters. An a posteriori error estimator for the DIRM method, which is based on a logarithmic Lipschitz constant for the nonlinearity, was also derived. This error estimator provides reliable information on the quality of the solution of the coupled system computed by the DIRM method. It can be efficiently calculated by determining the logarithmic norm of the Jacobi matrix of the nonlinearity using the successive constraint method. The presented numerical experiments demonstrate the usability of the proposed error estimator.

In the second part, a special partial integro-differential-algebraic equation was considered, which arises in simulation of electromagnetic devices coupled to electrical circuits. The distributed electromagnetic devices were modeled by Maxwell's equations in a magnetic vector potential formulation. First, we analysed magneto-quasistatic (MQS) field problems on a 2D domain. A spatial discretization by using the finite element method leads to a system of differential-algebraic equations (DAE). We studied the structural properties of the resulting system. In particular, we showed that this system has tractability index one. Furthermore, we investigated the passivity of the variational MQS problem and semidiscretized system by defining a storage function which describes the magnetic energy of the system. In the linear case, passivity was examined by means of positive realness of a transfer function. Based on this result, we developed a passivity-preserving balanced truncation model reduction method for linear MQS system. This method involves the solution of only one Lyapunov equation and provides a computable error bound. For model reduction of the nonlinear MQS system, we used proper orthogonal decomposition (POD) combined with the (matrix) discrete empirical interpolation method ((M)DEIM) for fast evaluation of the nonlinearity. Our model reduction approach is based on transforming the DAE system into an ODE form by exploiting a special block structure of the MQS model and applying standard model reduction methods to the resulting ODE system. It should be noted that the transformation to the ODE system requires the computation of an orthonormal basis of a certain subspace and, in general, destroys the sparsity of the system matrices. To overcome these computational difficulties, we used the underlying structure of the transformed system and performed all computations in terms of the original sparse data. For the POD reduced model, we proved the preservation of passivity, while for the POD-DEIM reduced model, we presented a passivity enforcement method based on perturbation of the output which depends on the errors introduced by DEIM. Numerical experiments for a single-phase 2D transformer demonstrate the performance of the presented model reduction methods.

Finally, MQS field problems on 3D domains were discussed. Here, a main difficulty is that the spatial discretization results in a singular DAE system. We presented a regularization method based on projecting out singular state components. For that purpose, we derived a condensed form for the system pencil which allows to decouple the nonlinear MQS system into the regular and singular part and to determine the subspaces corresponding to the infinite and zero generalized eigenvalues. This makes it possible to extend the balanced truncation and POD-DEIM model reduction methods to 3D linear and nonlinear MQS systems. We also studied the passivity of 3D MQS problems.

Future work will focus on improvements of model reduction algorithms for 3D linear and nonlinear MQS systems which can, for example, be achieved by using graph-theoretical algorithms for computing certain subspaces of involved incidence and loop matrices. Furthermore, the development of structure-preserving and energy-preserving model reduction methods for electromagnetic problems would be of great interest.

Bibliography

[AG01] M. Arnold and M. Günther. Preconditioned dynamic iteration for coupled differential-algebraic systems. *BIT*, 41(1):1–25, 2001.

[AH00] K. Afanasiev and M. Hinze. Adaptive control of a wake flow using proper orthogonal decomposition. *Lect. Notes Pure Appl. Math.*, 216:317–332, 2000.

[AH17] B.M. Afkham and J.S. Hesthaven. Structure preserving model reduction of parametric Hamiltonian systems. *SIAM J. Sci. Comput.*, 39(6):A2616–A2644, 2017.

[AHLS88] N. Aubry, P. Homes, J.L. Lumley, and E. Stone. The dynamics of coherent sructures in the wall region of a turbulent boundary layer. *J. Fluid Mech*, 11:115–173, 1988.

[Alt13] K. Altmann. *Numerische Verfahren der Optimalsteuerung von Magnetfeldern.* PhD thesis, Technische Universität Berlin, 2013.

[Ant05] A.C. Antoulas. *Approximation of Large-Scale Dynamical Systems.* SIAM, Philadelphia, PA, USA, 2005.

[Arn51] W.E. Arnoldi. The principle of minimized iterations in the solution of the matrix eigenvalue problem. *Quarterly Appl. Math.*, 9(17):17–29, 1951.

[ASG01] A.C. Antoulas, D.C. Sorensen, and S. Gugercin. A survey of model reduction methods for large-scale systems. *Contemporary Math.*, 280:193–219, 2001.

[Atk89] K. Atkinson. *An Introduction to Numerical Analysis.* Wiley, New York, USA, 1989.

[AV73] B.D.O. Anderson and S. Vongpanitlerd. *Network Analysis and Synthesis.* Prentice Hall, Englewood Cliffs, NJ, 1973.

[AV15] I. Anjam and J. Valdman. Fast MATLAB assembly of FEM matrices in 2D and 3D: Edge elements. *Appl. Math. Comput.*, 267(C):252–263, 2015.

[Bai02] Z. Bai. Krylov subspace techniques for reduced-order modeling of large-scale dynamical systems. *Appl. Numer. Math.*, 43:9–44, 2002.

[Bau12] S. Baumanns. *Coupled Electromagnetic Field/Circuit Simulation: Modeling and Numerical Analysis.* PhD thesis, Universität zu Köln, 2012.

[BB12] T. Breiten and P. Benner. Krylov-subspace based model reduction of nonlinear circuit models using bilinear and quadratic-linear approximation. *Progress in Industrial Mathematics at ECMI 2010, Mathematics in Industry*, 17:153–159, 2012.

[BBF14] U. Baur, P. Benner, and L. Feng. Model order reduction for linear and nonlinear systems: a system-theoretic perspective. *Arch. Comput. Methods Eng*, 21:331–358, 2014.

[BBS11] A. Bartel, S. Baumanns, and S. Schöps. Structural analysis of electrical circuits including magnetoquasistatic devices. *Appl. Numer. Math.*, 61(12):1257–1270, 2011.

[BCS12] A. Bartel, M. Clemens, and S. Schöps. Higher order half-explicit time integration of eddy current problems using domain substructuring. *IEEE Trans. Magn.*, 48(2):623–626, 2012.

[Ben07] G. Benderskaya. *Numerical Methods for Transient Field-Circuit Coupled Simulations Based on the Finite Integration Technique and a Mixed Circuit Formulation.* PhD thesis, Technische Universität Darmstadt, 2007.

[BG02] A. Bartel and M. Günther. A multirate W-method for electrical networks in state-space formulation. *J. Comput. Appl. Mathe.*, 147(2):411–425, 2002.

[BHtM11] P. Benner, M. Hinze, and E.J.W ter Maten, editors. *Model Reduction for Circuit Simulation*, volume 74 of *Lecture Notes in Electrical Engineering*. Springer-Verlag, Berlin, Heidelberg, 2011.

[BJWW00] H.T. Banks, M.L. Joyner, B. Winchesky, and W.P. Winfree. Nondestructive evaluaion using a reduced-order computational methodology. *Inverse Problems*, 16:1–17, 2000.

[BKS13a] P. Benner, P. Kürschner, and J. Saak. Efficient handling of complex shift parameters in the low-rank cholesky factor ADI method. *Numer. Algorithms*, 62(2):225–251, 2013.

[BKS13b] P. Benner, P. Kürschner, and J. Saak. An improved numerical method for balanced truncation for symmetric second order systems. *Math. Comput. Model. Dyn. Systems*, 19(6):593–615, 2013.

[BKS14] P. Benner, P. Kürschner, and J. Saak. Self-generating and efficient shift parameters in ADI methods for large Lyapunov and Sylvester equations. *Electron. Trans. Numer. Anal.*, 43:142–162, 2014.

[BLME07] B. Brogliato, R. Lozano, B. Maschke, and O. Egeland. *Dissipative Systems Analysis and Control: Theory and Applications.* Communications and Control Engineering. Springer-Verlag, London, 2007.

[BMS05] P. Benner, V. Mehrmann, and D. Sorensen, editors. *Dimension Reduction of Large-Scale Systems*, volume 45 of *Lecture Notes in Computational Science and Engineering*. Springer-Verlag, Berlin, Heidelberg, 2005.

[BOA$^+$17] P. Benner, M. Ohlberger, A.Patera, G. Rozza, and K. Urban, editors. *Model Reduction of Parametrized Systems*. SIAM, Philadelphia, 2017.

[Bos98] A. Bossavit. *Computational Electromagnetism*. Academic Press, San Diego, 1998.

[Bos01] A. Bossavit. "Stiff" problems in eddy-current theory and the regularization of Maxwell's equations. *IEEE Trans. Magn.*, 37(5):3542–3545, 2001.

[BRT17] T. Berger, T. Reis, and S. Trenn. Observability of linear differential-algebraic systems: A survey. In A. Ilchmann and T. Reis, editors, *Surveys in Differential-Algebraic Equations IV*, Differential-Algebraic Equations Forum, pages 161–219. Springer International Publishing, Cham, 2017.

[Brü10] T. Brül. *Dissipativity of Linear Quadratic Systems*. PhD thesis, Technische Universität Berlin, 2010.

[BS16] P. Benner and T. Stykel. Model reduction of differential-algebraic equations: a survey. In A. Ilchmann and T. Reis, editors, *Surveys in Differential-Algebraic Equations IV*, Differential-Algebraic Equations Forum. Springer-Verlag, Berlin, Heidelberg, 2016.

[Cam87] S.L. Campbell. A general form for solvable linear time varying singular systems of differential equations. *SIAM J. Math. Anal.*, 18(4):1101–1115, 1987.

[CC94] M.L. Crow and J.G. Chen. The multirate method for simulation of power system dynamics. *IEEE Trans. Power Syst.*, 9(3):1684–1690, 1994.

[Cia02] P. Ciarlet. *The Finite Element Method for Elliptic Problems*. SIAM, Philadelphia, PA, USA, 2002.

[CM95] Z.J. Cendes and J.B. Manges. A generalized tree-cotree gauge for magnetic field computation. *IEEE Trans. Magn.*, 31(3):1342–1347, 1995.

[CS05] W. Chen and M. Saif. Passivity and passivity based controller design of a class of switched control systems. *IFAC Proceedings Volumes*, 38(1):676–681, 2005. 16th IFAC World Congress.

[CS10] S. Chaturantabut and D.C. Sorensen. Nonlinear model reduction via discrete empirical interpolation. *SIAM J. Sci. Comput.*, 32(5):2737–2764, 2010.

[CS12] S. Chaturantabut and D.C. Sorensen. A state space error estimate for POD-DEIM nonlinear model reduction. *SIAM, J. Numer. Anal.*, 50(1):46–63, 2012.

[CSGB11] M. Clemens, S. Schöps, H. De Gersem, and A. Bartel. Decomposition and regularization of nonlinear anisotropic curl-curl DAEs. *COMPEL*, 30(6):1701–1714, 2011.

[CW02] M. Clemens and T. Weiland. Regularization of eddy current formulations using discrete grad-div operators. *IEEE Trans. Magn.*, 38(2):569–572, 2002.

[Dah59] G. Dahlquist. *Stability and error bounds in the numerical integration of ordinary differential equations.* Royal Institute of Technology, Stockholm, 1959.

[Dav66] E. Davison. A method for simplifying linear dynamical systems. *IEEE Trans. Automat. Control*, 11:93–101, 1966.

[Deo74] N. Deo. *Graph Theory with Applications to Engineering and Computer Science.* Prentice-Hall, Englewood Cliffs, N.J., 1974.

[DG16] Z. Drmač and S. Gugercin. A new selection operator for the discrete empirical interpolation method - improved a priori error bound and extensions. *SIAM J. Sci. Comput*, 38(2):A631–A648, 2016.

[Dir96] H. K. Dirks. Quasi-stationary fields for microelectronic applications. *Electrical Engineering*, 79(2):145–155, 1996.

[Ebe08] F. Ebert. *On Partitioned Simulation of Electrical Circuits using Dynamic Iteration Methods.* PhD thesis, Technische Universität Berlin, 2008.

[Enn84] D. Enns. Model reduction with balanced realization: an error bound and a frequency weighted generalization. *Proceedings of the 23rd IEEE Conference on Decision and Control (Las Vegas, 1984)*, pages 127–132, 1984.

[Eva98] L.C. Evans. *Partial differential equations.* Graduate studies in mathematics. American Mathematical Society, Providence, 1998.

[FPF01] C.A. Felippa, K.C. Park, and C. Farhat. Partitioned analysis of coupled mechanical systems. *Comput. Methods Appl. Mech. Engrg.*, 190:3247–3270, 2001.

[Fre04] R.W. Freund. SPRIM: structure-preserving reduced-order interconnect macromodeling. *Technical Digest of the 2004 IEEE/ACM International Conference on Computer-Aided Design*, pages 80–87, 2004.

[FRM08] F.D. Freitas, J. Rommes, and N. Martins. Gramian-based reduction method applied to large sparse power system descriptor models. *IEEE Trans. Power Systems*, 23(3):1258–1270, 2008.

[Fuk90] K. Fukunaga. *Introduction to Statistical Recognition*. Academic Press, New York, 1990.

[Glo84] K. Glover. All optimal hankel-norm approximations of linear multivariable systems and their L^∞-error bounds. *Internat. J. Control*, 39:1115–1193, 1984.

[Gri13] D.J. Griffiths. *Introduction to Electrodynamics*. Cambridge University Press, 2013.

[GSA03] S. Gugercin, D.C. Sorensen, and A.C. Antoulas. A modified low-rank Smith method for large-scale Lyapunov equations. *Numer. Algorithms*, 32:27–55, 2003.

[GSW13] S. Gugercin, T. Stykel, and S. Wyatt. Model reduction of descriptor systems by interpolatory projection methods. *SIAM J. Sci. Comput.*, 35(5):B1010–B1033, 2013.

[GTB+18] C. Gomes, C. Thule, D. Broman, P. Gorm Larsen, and H. Vangheluwe. Co-simulation: State of the art. *ACM Computing Surveys*, 51(3), 2018. Article No. 49.

[GW84] C.W. Gear and D.R. Wells. Multirate linear multistep methods. *BIT*, 24:484–502, 1984.

[HC13] T. Henneron and S. Clénet. Model order reduction of quasi-static problems based on POD and PGD approaches. *Eur. Phys. J. Appl. Phys.*, 64(2):24514:p1–p7, 2013.

[HC14] T. Henneron and S. Clénet. Model order reduction of non-linear magnetostatic problems based on POD and DEI methods. *IEEE Trans. Magn.*, 50(2):33–36, 2014.

[Hip00] R. Hiptmair. Multilevel gauging for edge elements. *Computing*, 64(2):97–122, 2000.

[HJ85] R.A. Horn and C.A. Johnson. *Matrix Analysis*. Cambrige University Press, 1985.

[HKBB+18] C. Hachtel, J. Kerler-Back, A. Bartel, M. Günther, and T. Stykel. Multirate DAE/ODE-simulation and model order reduction for coupled field-circuit systems. In U. Langer, W. Amrhein, and W. Zulehner, editors, *Scientific Computing in Electrical Engineering*, Mathematics in Industry. Springer International Publishing, Cham, 2018.

[HM80] D.J. Hill and P.J. Moylan. Dissipative dynamical systems: Basic input-output and state properties. *J. Franklin Inst.*, 309(5):327–357, 1980.

[HM89] H.A. Haus and J.R. Melcher. *Electromagnetic Fields and Energy*. Prentice Hall Books, 1989.

[HNW93] E. Hairer, S.P. Nørsett, and G. Wanner. *Solving Ordinary Differential Equations I - Nonstiff Problems*, volume 8 of *Springer Series in Computational Mathematics*. Springer-Verlag, Berlin, Heidelberg, 1993.

[HRB75] C.W. Ho, A.E. Ruehli, and P.A. Brennan. The modified nodal approach to network analysis. *IEEE Trans. Circuits Syst.*, 22:504 –509, 1975.

[HRSP07] D.B.P. Huynh, G. Rozza, S. Sen, and A.T. Patera. A successive constraint linear optimization method for lower bounds of parametric coercivity and inf-sup stability constants. *Comptes Rendus Mathematique*, 345(8):473–478, 2007.

[HSS08] M. Heinkenschloss, D.C. Sorensen, and K. Sun. Balanced truncation model reduction for a class of descriptor systems with application to the Oseen equations. *SIAM Sci. Comput.*, 30(2):1038–1063, 2008.

[HW96] E. Hairer and G. Wanner. *Solving Ordinary Differential Equations, Stiff and Differential-Algebraic Problems*. Springer-Verlag, Berlin, Heidelberg, 1996.

[HZ05] A. Horn and F. Zhang. Basic properties of the Schur complement. In F. Zhang, editor, *The Schur Complement and Its Applications*, volume 4 of *Numerical Methods and Algorithms*, pages 17–46. Springer Verlag, Boston, 2005.

[Ipa13] H. Ipach. Grafentheoretische Anwendung in der Analyse elektrischer Schaltkreise. Bachelor thesis, Universität Hamburg, 2013.

[IR98] K. Ito and S.S. Ravindran. A reduced basis method for control problems governed by PDEs. *Control and Estimation of Distributed Parameter Systems. Proceedings of the International Conference in Vorau, 1996*, pages 153–168, 1998.

[Jac99] J.D. Jackson. *Classical Electrodynamics*. Wiley, New York, NY, 1999.

[Jac02] J.D. Jackson. From Lorenz to Coulomb and other explicit gauge transformations. *Am. J. Phys.*, 70(9):917–928, 2002.

[JK96] Z. Jackiewicz and M. Kwapisz. Convergence of waveform relaxation methods for differential-algebraic systems. *SIAM J. Numer. Anal.*, 33(6):2303–2317, 1996.

[JR08] B. Jacob and T. Reis. Passivity criteria for infinite dimensional descriptor systems. *Proceedings in Applied Mathematics and Mechanics*, 8(1):10073–10076, 2008.

[JRS14] L. Jansen, U. Römer, and S. Schöps. Abstract differential-algebraic equations in electrical engineering. In *85th Annual Meeting of the International Association of Applied Mathematics and Mechanics (GAMM 2014)*, Erlangen-Nürnberg, 2014.

[KBS17] J. Kerler-Back and T. Stykel. Model reduction for linear and nonlinear magneto-quasistatic equations. *Int. J. Numer. Meth. Eng.*, 111(13):1274–1299, 2017.

[KM06] P. Kunkel and V. Mehrmann. *Differential-Algebraic Equations. Analysis and Numerical Solution.* EMS Publishing House, 2006.

[Kow06] P. Kowal. Null space of a sparse matrix. `http://www.mathworks.co.uk/matlabcentral/fileexchange/11120`, 2006.

[KS14] J. Kerler and T. Stykel. Model order reduction and dynamic iteration for coupled systems. In *Proceedings of the GAMM Annual Meeting (GAMM 2014, Erlangen, March 10-14, 2014)*, pages 527–528, 2014.

[KS15] J. Kerler and T. Stykel. Model order reduction for magneto-quasistatic equations. In *IFAC-PapersOnLine, Proceedings of the 8-th Vienna International Conference on Mathematical Modelling (MATHMOD 2015, Vienna, Austria, February 18-20, 2015)*, volume 48, pages 240–241, 2015.

[KV02] K. Kunisch and S. Volkwein. Galerkin proper orthogonal decomposition methods for a general equation in fluid dynamics. *SIAM J. Numer. Anal.*, 40:492–515, 2002.

[LBH84] R.W. Lewis, P. Bettess, and E. Hinton, editors. *Numerical methods in coupled systems.* Wiley, Chichester New York, 1984.

[LHPW87] A.J. Laub, M.T. Heath, C.C. Paige, and R.C. Ward. Computation of system balancing transformations and other applications of simultaneous diagonalization algorithms. *IEEE Trans. Automat. Control*, 32:115–122, 1987.

[LMT13] R. Lamour, R. März, and C. Tischendorf. *Differential-Algebraic Equations. Analysis and Numerical Solutions: A Projector Bases Analysis.* Differential-Algebraic Equations Forum. Springer-Verlag, Berlin, Heidelberg, 2013.

[LS01] W. Liu and V. Sreeram. Model reduction for singular systems. *Internat. J. Syst. Sci.*, 32:1205–1215, 2001.

[LT01] H.V. Ly and H.T. Tran. Modelling and control of physical processes using proper orthogonal decomposition. *Math. and Comput. Modeling*, 33:223–236, 2001.

[LW02] J.-R. Li and J. White. Low rank solution of Lyapunov equations. *SIAM J. Matrix Anal. Appl.*, 24(1):260–280, 2002.

[Mar66] S. Marschall. An approximate method for reducing the order of a linear system. *Contr. Eng.*, 10:642–648, 1966.

[Max65] J.C. Maxwell. A dynamical theory of the electromagnetic field. *Philosophical Transactions of the Royal Society of London*, 155(1):459–513, 1865.

[Meh15] V. Mehrmann. Index concepts for differential-algebraic equations. In B. Engquist, editor, *Encyclopedia Applied and Computational Mathematics*, pages 676–681. Springer-Verlag, Berlin, Heidelberg, 2015.

[MHCG16] L. Montier, T. Henneron, S. Clénet, and B. Goursaud. Transient simulation of an electrical rotating machine achieved through model order reduction. *Adv. Model. Simul. Eng. Sci.*, 3:10, 2016.

[MHCG17] L. Montier, T. Henneron, S. Cleńet, and B. Goursaud. Balanced proper orthogonal decomposition applied to magnetoquasistatic problems through a stabilization methodology. *IEEE Trans. Magn.*, 53(3):1–10, 2017.

[MN92] U. Miekkala and O. Nevanlinna. Convergence of dynamic iteration methods for initial value problems. *Numer. Funct. Anal. Optim.*, 13:203–221, 1992.

[Mon03] P. Monk. *Finite Element Methods for Maxwell's Equations*. Numerical Mathematics and Scientific Computation. Oxford University Press, 2003.

[Moo81] B.C. Moore. Principal component analysis in linear systems: controllability, observability, and model reduction. *IEEE Trans. Automat. Control*, AC-26(1):17–32, 1981.

[Mun02] I. Munteanu. Tree-cotree condensation properties. *ICS Newsletter (International Compumag Society)*, 9:10–14, 2002.

[Néd80] J.C. Nédélec. Mixed finite elements in $\mathbb{R}^3$. *Numerische Mathematik*, 35(3):315–341, Sep 1980.

[NST14] S. Nicaise, S. Stingelin, and F. Tröltzsch. On two optimal control problems for magnetic fields. *Comput. Methods Appl. Math.*, 14(4):555–573, 2014.

[NT13] S. Nicaise and F. Tröltzsch. A coupled Maxwell integrodifferential model for magnetization processes. *Mathematische Nachrichten*, 287(4):432–452, 2013.

[OCP98] A. Odabasioglu, M. Celik, and L.T. Pileggi. PRIMA: Passive reduced-order interconnect macromodeling algorithm. *IEEE Trans. Circuits Syst.*, 17(8):645–654, 1998.

[Pan88] C.C. Pantelides. The consistent initialization of differential-algebraic systems. *SIAM J. Sci. Statist. Comput.*, 9:213–231, 1988.

[Pec04] C. Pechstein. Multigrid-Newton-methods for nonlinear magnetostatic problems. Diploma thesis, Johannes Kepler Universität Linz, February 2004.

[Pen00] T. Penzl. A cyclic low-rank Smith method for large sparse Lyapunov equations. *SIAM J. Sci. Comput.*, 24(4):1401–1418, 1999/2000.

[PM16] L. Peng and K. Mohsen. Symplectic model reduction of Hamiltonian systems. *SIAM J. Sci. Comput.*, 38(1):A1–A27, 2016.

[Pol98] J.G. Polushin. Further properties of nonlinear quasipassive systems. *In Proceedings of the 37th IEEE conference on Decision and Control (Tampa, Florida, USA, December 1998), IEEE*, pages 4144–4149, 1998.

[PS82] L. Pernebo and L. Silverman. Model reduction via balanced state space representations. *IEEE Trans. Automat. Control*, 27(2):382–387, 1982.

[PT18] J. Pade and C. Tischendorf. Waveform relaxation: a convergence criterion for differential-algebraic equations. *Numer. Algorithms*, 2018. DOI: 10.1007/s11075-018-0645-5.

[QV94] A. Quarteroni and A. Valli. *Numerical Approximation of Partial Differential Equations.* Springer-Verlag, Berlin Heidelberg, 1994.

[Ral81] L.B. Rall. *Automatic Differentiation. Techniques and Applications.* Springer-Verlag, Berlin Heidelberg, 1981.

[Rei14] T. Reis. Mathematical modeling and analysis of nonlinear time-invariant RLC circuits. In P Benner, R. Findeisen, D. Flockerzi, U. Reichl, and K. Sundmacher, editors, *Large-Scale Networks in Engineering and Life Sciences*, Modeling and Simulation in Science, Engineering and Technology, pages 125–198. Birkhäuser, Cham, 2014.

[RF94] D. Rempfer and H.F. Fasel. Dynamics of three-dymensional coherent structures in a flat-plate boundary layer. *J. Fluid. Mech.*, 275:257–283, 1994.

[Rhe84] W.C. Rheinboldt. Differential-algebraic systems as differential equations on manifolds. *Math. Comput.*, 43:473–482, 1984.

[Ria08] R. Riaza. *Differential Algebraic Systems. Analytical Aspects and Circuit Applications.* World Scientific, River Edge, NJ, USA, 2008.

[RP02] M. Rathinam and L. Petzold. Dynamic iteration using reduced order models: a method for simulation of large scale modular systems. *SIAM J. Numer. Anal.*, 40(4):1446–1474, 2002.

[RS07] T. Reis and T. Stykel. Stability analysis and model order reduction of coupled systems. *Math. Comput. Model. Dyn. Syst.*, 13(5):413–436, 2007.

[RS10] T. Reis and T. Stykel. PABTEC: passivity-preserving balanced truncation for electrical circuits. *IEEE Trans. CAD Integr. Circuits Syst.*, 29(9):1354–1367, 2010.

[RS11] T. Reis and T. Stykel. Lyapunov balancing for passivity-preserving model reduction of RC circuits. *SIAM J. Appl. Dyn. Syst.*, 10(1):1–34, 2011.

[RW06] M.J. Rewieński and J. White. Model order reduction for nonlinear dynamical systems based on trajectory piecewise-linear approximations. *Linear Algebra Appl.*, 415(2-3):426–454, 2006.

[SA12] T. Schierz and M. Arnold. Stabilized overlapping modular time integration of coupled differential-algebraic equations. *Appl. Numer. Math.*, 62:1491–1502, 2012.

[Sch03a] A. Schneebeli. An $H(curl;\Omega)$-conforming FEM: Nédélec's elements of the first type. `https://www.dealii.org/reports/nedelec/nedelec.pdf`, 2003.

[Sch03b] S. Schulz. Four lectures on differential-algebraic equations. `https://www.mathematik.hu-berlin.de/~steffen/pub/introduction_to_daes_497.pdf`, 2003.

[Sch11] S. Schöps. *Multiscale modeling and multirate time-integration of field/circuit coupled problems.* PhD thesis, Bergische Universität Wuppertal, 2011.

[SGW13] S. Schöps, H. De Gersem, and T. Weiland. Winding functions in transient magnetoquasistatic field-circuit coupled simulations. *COMPEL*, 32(6):2063–2083, 2013.

[Sir87] L. Sirovich. Turbulence and the dynamics of coherent structures, parts I-III. *Quart. Appl. Math.*, XLV:561–590, 1987.

[SK98] S.Y. Shvartsman and Y. Kevrikidis. Nonlinear model reduction for control of distributed parameter systems: a computer-assisted study. *American Institute of Chemical Engineers (AIChE) Journal*, 44:1579–1595, 1998.

[SLL15] B. Schweizer, P. Li, and D. Lu. Explicit and implicit cosimulation methods: Stability and convergence analysis for different solver coupling approaches. *J. Comput. Nonlinear Dyn.*, 10(5):051007, 2015.

[Smi85] G.D. Smith. *Numerical Solution of Partial Differential Equations: Finite Difference Methods.* Oxford University Press, 1985.

[Söd06] G. Söderlind. The logarithmic norm. History and modern theory. *BIT, Numer. Math.*, 46:631–652, 2006.

[SS12] T. Stykel and V. Simoncini. Krylov subspace methods for projected Lyapunov equations. *Appl. Numer. Math.*, 62:35–50, 2012.

[Sty04] T. Stykel. Gramian-based model reduction for descriptor systems. *Math. Control Signal Systems*, 16:297–319, 2004.

[Sty08] T. Stykel. Low-rank iterative methods for projected generalized Lyapunov equations. *Electron. Trans. Numer. Anal.*, 30:187–202, 2008.

[Sty11] T. Stykel. Balancing-related model reduction of circuit equations using topological structure. In P. Benner, H. Hinze, and E.J.W. ter Maten, editors, *Model Reduction for Circuit Simulation*, volume 74 of *Lecture Notes in Electrical Engineering*, pages 53–80. Springer-Verlag, Berlin, Heidelberg, 2011.

[SV18] J. Saak and M. Voigt. Model reduction of constrained mechanical systems in M-M.E.S.S. In *IFAC-PapersOnLine, Proceedings ot the 9-th Vienna International Conference on Mathematical Modelling (MATHMOD 2018, Vienna, Austria, February 21-23, 2018)*, volume 51, pages 661–666, 2018.

[SvdVR08] W.H.A. Schilders, H.A. van der Vorst, and J. Rommes, editors. *Model Order Reduction: Theory, Research Aspects and Applications*, volume 13 of *Mathematics in Industry*. Springer-Verlag, Berlin, Heidelberg, 2008.

[SW13] T. Stykel and C. Willbold. Model reduction based optimal control for field-flow fractionation. In *Proceedings of the GAMM Annual Meeting (GAMM 2013, Novi Sad, March 18-22, 2013)*, 2013.

[Tis03] C. Tischendorf. *Coupled systems of differential algebraic and partial differential equations in circuit and device simulation.* Habilitation thesis, Humboldt-Universität zu Berlin, 2003.

[TP87] M.S. Tombs and I. Postlethwaite. Truncated balanced realization of a stable non-minimal state-space system. *International Journal of Control*, 46(2):1319–1330, 1987.

[USKB12] M.M. Uddin, J. Saak, B. Kranz, and P. Benner. Computation of a compact state space model for an adaptive spindle head configuration with piezo actuators using balanced truncation. *Prod. Eng. Res. Devel*, 6(6):577–586, 2012.

[vdS00] A. van der Schaft. *L2-Gain and Passivity Techniques in Nonlinear Control.* Springer-Verlag, Basel, 2000.

[Vol99] S. Volkwein. Proper orthogonal decomposition and singular value decomposition. Technical report, Graz University, 1999.

[Vol13] S. Volkwein. Proper orthogonal decomposition: Theory and reduced-order modelling. Lecture Notes. University of Konstanz, 2013.

[VS94] J. Vlach and K. Singhal. *Computer Methods for Circuit Analysis and Design.* Van Nostrand Reinhold, New York, 1994.

[Wac13] E. Wachspress. *The ADI Model Problem.* Springer-Verlag, New York, 2013.

[Wei77] T. Weiland. A discretization method for the solution of Maxwell's equations for six-component fields. *Electron. Commun.*, 31(3):116–120, 1977.

[Wil72] J.C. Williams. Dissipative dynamical systems. Part I: General theory and Part II: Linear systems with quadric supply rates. *Arch. Rational Mechanics Analysis*, 45(5):321–393, 1972.

[Wil16] C. Willbold. *Model reduction and optimal control in field-flow fractionation.* PhD thesis, Universität Augsburg, 2016.

[WP99] M. Witting and T. Pröpper. Cosimulation of electromagnetic fields and electrical networks in the time domain. *Surveys Math. Indust.*, 9:101–116, 1999.

[WR00] J.A. Weideman and S.C. Reddy. A MATLAB Differentiation Matrix Suite. *ACM Trans. Math. Softw.*, 26(4):465–519, December 2000.

[WSH14] D. Wirtz, D.C. Sorensen, and B. Haasdonk. A posteori error estimation for DEIM reduced nonlinear dynamical systems. *SIAM J. Sci. Comput.*, 36(2):A311–A338, 2014.

[Zei90a] E. Zeidler. *Nonlinear Functional Analysis and Its Applications. II/A Linear Monotone Operators.* Springer, New York, 1990.

[Zei90b] E. Zeidler. *Nonlinear Functional Analysis and Its Applications. II/B Nonlinear Monotone Operators.* Springer, New York, 1990.

[Zul08] W. Zulehner. *Numerische Mathematik. Eine Einführung anhand von Differentialgleichungsproblemen Band 1: Stationäre Probleme.* Birkhäuser Basel, 2008.

[Zul11] W. Zulehner. *Numerische Mathematik. Eine Einführung anhand von Differentialgleichungsproblemen Band 2: Instationäre Probleme.* Birkhäuser Basel, 2011.